T0383546

Towards a Common Software/Hardware Methodology for Future Advanced Driver Assistance Systems

The DESERVE Approach

RIVER PUBLISHERS SERIES IN TRANSPORT TECHNOLOGY

Indexing: All books published in this series are submitted to Thomson Reuters Book Citation Index (BkCI), CrossRef and to Google Scholar.

The "River Publishers Series in Transport Technology" is a series of comprehensive academic and professional books which focus on theory and applications in the various disciplines within Transport Technology, namely Automotive and Aerospace. The series will serve as a multi-disciplinary resource linking Transport Technology with society. The book series fulfils the rapidly growing worldwide interest in these areas.

Books published in the series include research monographs, edited volumes, handbooks and textbooks. The books provide professionals, researchers, educators, and advanced students in the field with an invaluable insight into the latest research and developments.

Topics covered in the series include, but are by no means restricted to the following:

- Automotive
- Aerodynamics
- Aerospace Engineering
- Aeronautics
- Multifunctional Materials
- Structural Mechanics

For a list of other books in this series, visit www.riverpublishers.com

Towards a Common Software/Hardware Methodology for Future Advanced Driver Assistance Systems

The DESERVE Approach

Editors

Guillermo Payá-Vayá

Leibniz Universität Hannover
Germany

Holger Blume

Leibniz Universität Hannover
Germany

River Publishers

Published, sold and distributed by:
River Publishers
Alsbjergvej 10
9260 Gistrup
Denmark

River Publishers
Lange Geer 44
2611 PW Delft
The Netherlands

Tel.: +45369953197
www.riverpublishers.com

ISBN: 978-87-93519-14-5 (Hardback)
978-87-93519-13-8 (Ebook)

Printed on acid-free paper.

Contents

Preface

The European research project DESERVE (DEvelopment platform for Safe and Efficient dRiVE, 2012–2015) had the aim of designing and developing a platform tool to cope with the continuously increasing complexity and the simultaneous need to reduce costs for future embedded Advanced Driver Assistance Systems (ADAS). For this purpose, the DESERVE platform profits from cross-domain software reuse, standardization of automotive software component interfaces, and easy but safety-compliant integration of heterogeneous modules. This enables the development of a new generation of ADAS applications, which challengingly combine different functions, sensors, actuators, hardware platforms, and Human Machine Interfaces (HMI).

This book provides a detailed overview of the different research activities conducted in the course of the DESERVE project. After introducing the aims of the DESERVE project in Chapter 1, selected achievements of the DESERVE project are presented in three different parts. Part I is dedicated to the ADAS development platform developed during the DESERVE project.

- Chapter 2 covers the methodology and concepts that are part of the generic DESERVE platform as the basis and key enabler for the development of new assistance systems. It describes the entire spectrum of aspects, e.g., modularity, interfaces, and standards, to be considered for the use of the DESERVE platform.
- Chapter 3 describes the development of realistic models for driver behavior as part of the DESERVE tool-chain needed for the evaluation of complex ADAS systems and driver-vehicle-environment interactions. The modelling system was used to simulate two different driving scenarios.
- Chapter 4 presents component based middleware, e.g., RTMaps and ADTF, for supporting the developer of complex systems with typical challenges like multi-sensor support, synchronization issues, and modularity. By means of different exemplary applications, in which modules like simulators or prototyping systems are connected to the middleware, the flexibility of the DESERVE tool-chain is demonstrated.

- Chapter 5 describes a model-in-the-loop approach for tuning ADAS parameters. Using the AVL CAMEO tool, model-based design space exploration and validation of a complex ADAS function is performed.

In Part II, ADAS applications used as test functions in the DESERVE project are explained.

- Chapter 6 presents an application of deep-learning techniques for semantic segmentation of camera images (i.e., Scene Labeling). After explaining the algorithmic basics, an FPGA-based implementation is presented and evaluated.
- Chapter 7 covers a system coupling an FPGA-based signal processing architecture for MIMO radar with a PC-based ADTF data post-processing. The hardware-software combination maximizes processing performance and minimizes development time of complex systems.
- Chapter 8 describes a design space exploration for online calibration of wide baseline stereo camera systems using sparse feature correspondences in stereo images. Challenges in hardware implementations of feature matching are presented and hardware-specific solutions are discussed.
- Chapter 9 presents a first approach of arbitration and sharing vehicle control between driver and assistance system based on modelling vehicles and driver behavior and intentions. Fuzzy logic techniques are used to implement the control sharing and simulations allow testing of the systems.

Part III covers the validation and evaluation of two exemplary applications of the DESERVE platform.

- Chapter 10 aims at exploring effective design of Human Machine Interface (HMI). During the DESERVE project, in-vehicle HMI solutions for different functions were developed. The HMI design process for an exemplary function is described in this chapter.
- Chapter 11 shows a prototype system for vehicle-in-the-loop testing of ADAS functions that additionally analyzes the energy efficiency of the prototyped system. Combined with multi-sensor simulation, a virtual environment for testing ADAS functions is provided.

Further detailed information about the contributions of DESERVE can be found in the list of project deliverables referenced in each chapter.

This work was supported by the European Commission under the Artemis Joint Undertaking in the scope of the DESERVE project. We would like to thank all authors and co-authors for their excellent contributions. Special thanks to Matti Kutila for the efficient managing of the complete DESERVE project over three years. Further thanks to Martin Kunert who well-coordinated subprojects and who actively supported our work. Furthermore we want to thank the River Publishers Team, in particularly Mr. Mark de Jongh and Ms. Junko Nakajima for their great support.

We hope that you will enjoy reading this book.

Guillermo Payá Vayá
Holger Blume

March 22th, 2017
Hannover (Germany)

List of Contributors

Abhishek Ravi, *AVL List Gmbh, Austria*

André Rolfsmeier, *dSpace GmbH, Germany*

Andrea Saroldi, *C.R.F. S.C.p.A , Italy*

Angel Cuevas, *CTAG – Centro Tecnológico de Automoción de Galicia, Spain*

Caterina Calefato, *Unimore – University of Modena and Reggio Emilia – Italy*

Chiara Ferrarini, *Unimore – University of Modena and Reggio Emilia – Italy*

Clément Galko, *Univ. Rouen, UNIROUEN, ESIGELEC, IRSEEM 76000 Rouen, France*

David González, *INRIA, France*

Elisa Landini, *RE:Lab srl, Italy*

Eugen Schubert, *Advanced Engineering Sensor Systems, Robert Bosch GmbH, Leonberg, Germany*

Eva M. García Quinteiro, *CTAG – Centro Tecnológico de Automoción de Galicia, Spain*

Fabio Tango, *CRF – Centro Ricerche Fiat, Italy*

Fawzi Nashashibi, *INRIA, France*

Florian Giesemann, *Institute of Microelectronic Systems, Leibniz Universität Hannover, Hannover, Germany*

Frank Badstübner, *Infineon Technologies AG, Germany*

Frank Meinl, *Advanced Engineering Sensor Systems, Robert Bosch GmbH, Leonberg, Germany*

Gideon Reade, *ASL, U.K.*

Guillermo Payá-Vayá, *Institute of Microelectronic Systems, Leibniz Universität Hannover, Hannover, Germany*

Gwenaël Dunand, *Intempora, France*

Hans Michael Koegeler, *AVL List Gmbh, Austria*

Hariharan Narasimman, *Univ. Rouen, UNIROUEN, ESIGELEC, IRSEEM 76000 Rouen, France*

Holger Blume, *Institute of Microelectronic Systems, Leibniz Universität Hannover, Hannover, Germany*

Jens Klimke, *Institute for Automotive Engineering, RWTH Aachen University, Steinbachstraße 7, 52074 Aachen, Germany*

Joshué Pérez, *INRIA, France*

Lars Krüger, *Daimler AG, Vision Enhancement, Ulm, Germany*

Lutz Eckstein, *Institute for Automotive Engineering, RWTH Aachen University, Steinbachstraße 7, 52074 Aachen, Germany*

Marga Sáez Tort, *CTAG – Centro Tecnológico de Automoción de Galicia, Spain*

Martin Kunert, *Advanced Engineering Sensor Systems, Robert Bosch GmbH, Leonberg, Germany*

Matthias Limmer, *Vision Enhancement, Daimler AG, Germany*

Matti Kutila, *VTT Technical Research Center of Finland Ltd., Finland*

Nereo Pallaro, *Centro Ricerche Fiat, Italy*

Nico Mentzer, *Institute of Microelectronic Systems, Leibniz Universität Hannover, Hannover, Germany*

Nora von Egloffstein, *Daimler AG, Vision Enhancement, Ulm, Germany*

Ralf Ködel, *Infineon Technologies AG, Germany*

Roberto Montanari, *RE:Lab srl, Italy*

Romain Rossi, *Univ. Rouen, UNIROUEN, ESIGELEC, IRSEEM 76000 Rouen, France*

Vicente Milanés, *INRIA, France*

Werner R. Ritter, *Vision Enhancement, Daimler AG, Germany*

Wilhelm Maurer, *Infineon Technologies AG, Germany*

Xavier Savatier, *Univ. Rouen, UNIROUEN, ESIGELEC, IRSEEM 76000 Rouen, France*

List of Figures

List of Tables

List of Abbreviations

ABS	Anti-lock Breaking System
ACC	Adaptive Cruise Control
ADAS	Advanced Driver Assistance Systems
ADC	Analog-to-digital converter
AEB	Automatic/Autonomous Emergency Braking
AR	Autoregressive
ASIC	Application-Specific Integrated Circuit
ASIP	Application-Specific Instruction-Set Processor
avg	Average
BASt	German Federal Highway Research Institute
bpp	Bit per pixel
BRIEF	Binary Robust Independent Elementary Features
BRISK	Binary Robust Invariant Scalable Keypoints
BSD	Blind Spot Detection
CA-CFAR	Cell-averaging constant false alarm rate
CAN Bus	Controller Area Network
CDMA	Code division multiple access
CenSurE	Center Surround Extremas
CFAR	Constant false alarm rate
CM4SL	Carmaker for simulink
CMbB	Collision Mitigation by Braking
CMOS	Complementary Metal-Oxide-Semiconductor
CNN	Convolutional Neural Network
COR	Customized Output Range
CPU	Central Processing Unit
CRF	Conditional Random Field
CUT	Cell under test
DAISY	Name of a feature descriptor
DAS	Driver assistance systems
DBC	data base CAN
DIF	Decimation-in-frequency

DMA	driving monitoring automotive
DOA	Direction of arrival
DoE	Design of Experiment
DoG	Difference of Gaussian
DRAM	Dynamic random-access memory
ECU	Electronic Control Unit
ESC	Electronic Stability Control
ESPRIT	Estimation of signal parameters via rotational invariant techniques
FAST	Features from Accelerated Segment Test
FCW	Frontal Collision Warning or Forward Collision Warning
FDM	Frequency-division multiplexing
FFT	Fast Fourier transform
FIR	Finite impulse response
FMCW	Frequency-modulated continuous-wave
FN(R)	False Negative (Rate)
FP(R)	False Positive (Rate)
FPGA	Field-Programmable Gate Array
fps	Frames per second
FREAK	Fast Retina Keypoint
GB	Geometry-based
GOPS	Billion Operations Per Second
GPGPU	General Purpose Graphics Processing Unit
GPP	General Purpose Processor
GPU	Graphics Processing Unit
HD	High-definition, 1280×720 pixel
HiL	Hardware in the Loop
HMI	Human-machine interface
HW	Hardware
I/O	input/output
I2C	Inter-Integrated Circuit
IMU	Inertial measurement unit
IU	Intersection over Union
IWI	information-warning-intervention
KD-Tree	K-dimensional tree
KPI	Key Performance Indicator
LCA	Lane Change Assistant
LDW	Lane Departure Warning

LKA	Lane Keeping Assistance
LoG	Laplacian of Gaussian
LSB	Least significant bit
LUT	Lookup table
MCC	Matthews Correlation Coefficient
MDC	Multi-path delay commutator
MiL	Model in the Loop
MIMO	Multiple-input multiple-output
MLP	Multi Layer Perceptron
MOPS	Million Operations Per Second
MUSIC	Multiple signal classification
NCI	Non-coherent integration
NHTSA	National Highway Traffic Safety Administration
NMEA	National Marine Electronics Association
NNB	Nearest-Neighbor-Based
NNDR	Nearest-Neighbor Distance Ratio
OpenCL	Open Computing Language
OpenGL	Open Graphics Library
ORB	Oriented FAST and Rotated BRIEF
OS-CFAR	Ordered-statistic constant false alarm rate
PCA	Principal Component Analysis
PID	proportional, integral, derivative controller
QVGA	Quarter Video Graphics Array, 320×240 pixel
RCS	Radar cross-section
RDE	Reak Driving Emissions
ReLU	Rectifier Linear Unit
RMS	Root Mean Square
RPM	Revolution per minute
RTSP	Real Time Streaming Protocol
SAE	Society of Automotive Engineers
SDF	Single-path delay feedback
SIFT	Scale-Invariant Feature Transform
SIP	Session Initialization Protocol
SLA	Speed Limit Assistant
SNR	Signal-to-noise ratio
SoP	Start of Production
SQNR	Signal-to-quantization-noise ratio
SRAM	Static random-access memory

SURF	Speeded Up Robust Features
SW	Software
TB	Threshold-Based
TDM	Time-division multiplexing
TP	True Positive
UUT	Unit Under Test
VGA	Video Graphics Array, 640×480 pixel

1

The DESERVE Project: Towards Future ADAS Functions

Matti Kutila[1] and Nereo Pallaro[2]

[1]VTT Technical Research Center of Finland Ltd., Finland
[2]Centro Ricerche Fiat, Italy

1.1 Project Aim

This book aims to outline the major innovations introduced by the DESERVE (DEvelopment platform for Safe and Efficient dRiVe) project. The project started in September 2012 and finished on February 2015 after 3,5 years heavy working and was coordinated by VTT Technical Research Centre of Finland Ltd. The project was co-funded by the European Commission under the ECSEL EU-Horizon 2020 programme. The project was a joint effort of major vehicle manufacturers (Volvo, Daimler, Fiat), component suppliers (Continental, Ficosa, AVL, Bosch, NXP, Infineon, dSPACE, ASL Vision, Ramboll, TTS, Technolution), research institutes (VTT, ICOOR, ReLab, INRIA, CTAG) and universities (VisLab, IRSEEM, ARMENIS, IKA, INTEMPORA, Leibniz Universität Hannover).

VISION

DESERVE will design and build an ARTEMIS Tool Platform based on the standardisation of the interfaces, software (SW) reuse, development of common non-competitive SW modules, and easy and safety-compliant integration of standardised hardware (HW) or SW from different suppliers

The main research question was to identify the optimal sensor solutions for the DESERVE platform which are required by the selected ADAS functions for supporting transition to automated vehicles. 22 different modules were selected to be implemented to 11 driver support applications according to user needs when starting development process:

- Lane change assistance system
- Pedestrian safety systems
- Forward/rearward looking system (distant range)
- Adaptive light control
- Park assistance
- Night vision system
- Cruise control system
- Traffic sign and traffic light recognition
- Map-supported systems
- Vehicle interior observation
- Driver monitoring

The project created the methodology framework for integrating embedded hardware and software modules was created which enables better interoperability of automotive industry products and third party aftersales components. This approach is also beneficial to comprise the problem for guaranteeing safety and security problems when new components are added to the complex software and hardware stacks.

The initial project objective has been defined in the Table 1.1 with having measurable verification of the expected results.

Table 1.1 Scientific and technical objectives

Scientific and Technical Objectives	Measurable and Verifiable Form
The definition and implementation of a model-driven process for the compositional development of safety critical systems that allows the smooth integration of existing components and functions in a new framework.	By defining an analysis methodology to establish an industrially applicable process for exploration of design spaces and multi-criteria constraint satisfaction, with particular regard to safety properties. **Verification: 90% or more of the applications identified could be developed with the proposed platform.**
The development of an innovative embedded vehicle platform capable of supporting the fast and reliable development of ADAS and efficient Eco-driving functions.	By implementing demonstrators for active and passive safety of drivers and all road users in the three macro-areas in the automotive domain such as:

Table 1.1 Continued

	• Technical, safety and efficiency impact assessment of resulting prototypes following the evaluation methodologies identified in project PREVAL and in line with INTERACTIVE evaluation methodologies. • Cost-Benefits analysis. • Evaluation of cost reduction in comparison with conventional Driver Assistance Systems. **Verification: 90% or more of the developed applications showed more than 15% of reduction in development time and cost.**
The integration of existing vehicle sensors and actuators in a unified SW framework for multiple safety and Eco-driving applications.	Existence of a cost-effective and flexible SW platform, able to be used with available sensors/actuators. **Verification: 90% or more of the developed applications show more than 15% reduction in development duration and cost.**
The adaptation of the current data fusion, HMI and driver's behaviour modules to provide suitable and harmonised middleware for the different safety and Eco-driving functions.	By applying the V-model and developing high level services and Application Protocol Interface (API) that can be used in a wide range of safety-related use cases. Via multi-modal HMI with user related and driver behaviour assessment through tests in driving simulator and in prototype vehicles. **Verification: Statistical evidence of improvement of driver acceptance between existing (on the market) and DESERVE-developed functions. Subjective evaluation through questionnaires.**
The implementation of a new method and relative tools for ADAS functions development.	Existence of new tools for development of Driver Assistance Systems, including data fusion visualisation, algorithm development, actuation simulation, etc. **Verification: Evidence that the method is suitable for effective ADAS developments:** • **Results of the test case development** • **Results of workshops with main stakeholders, OEMs and automotive suppliers.**

The developed applications are tested and validated in different demonstration vehicles for showing that DESERVE methodology is not limited to one single vehicle type. The project demonstration vehicles are:

- two medium class passenger cars from Fiat
- research passenger car from VTT
- luxury passenger car from Daimler
- heavy goods vehicle from Volvo
- driver training truck from TTS

Additionally, tests will also be conducted in simulators, e.g. a simulator for driver monitoring functions and a simulator for cruise control systems.

1.2 Project Structure

The project was divided into 8 sub-projects (see Figure 1.1) in order to keep the whole development chain manageable and taking different automotive orientated technical challenges into account.

This project workflow also enabled professional development process starting from the requirements and finishing to the validation phase. One sub-project was engaged with specifying and designing the DESERVE platform and three sub-projects for doing implementation.

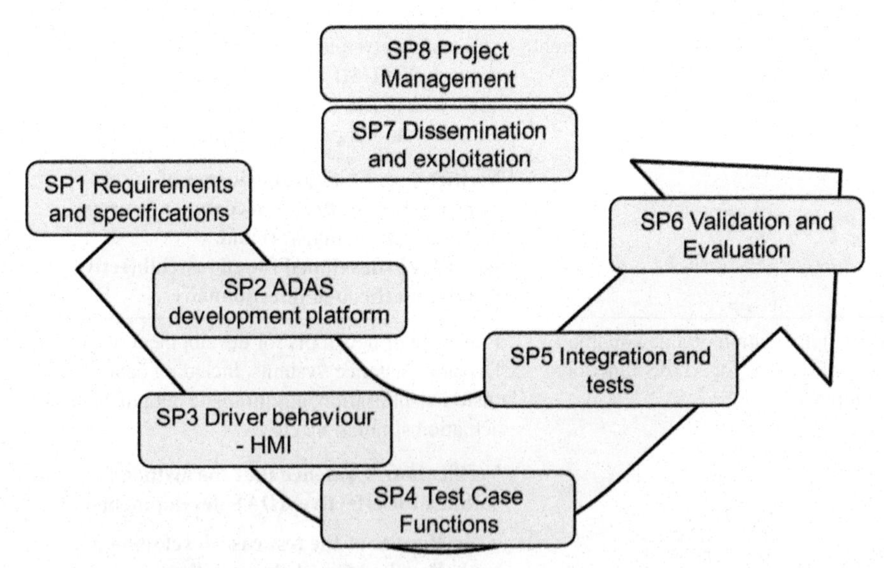

Figure 1.1 The DESERVE V-shape development process.

1.3 DESERVE Platform Design

The project developed the framework methodology (see Figure 1.2) to integrate new software components to car environment. In practise, the methodology verified with implementing two alternative solutions which were adapted to fit to the project framework design. The one bases on ADTF which is mainly utilised by the German automotive industry and RTMaps which is implemented by the other demonstrators. Since the aim is to introduce a solution which will be exploited in real vehicles both solutions this gives good bases to bring the specified framework to cars in future within next 5 years.

1.4 The Project Innovation Summary

The project was not limited to the framework design but was also further developing the current in-vehicle technology. The specific areas where steps were taken forward are:

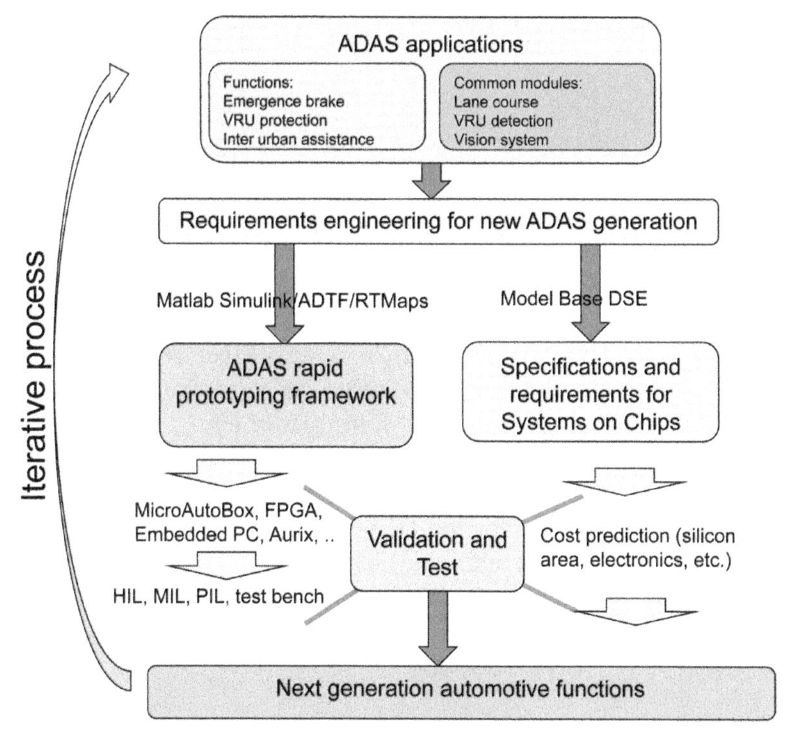

Figure 1.2 DESERVE platform concept for speeding up the ADAS function development time.

- Night time environment perception
- Driver monitoring topics: Drowsiness and distraction detection
- Embedded in-vehicle computing system: Setting up FPGA based automotive CPUs
- Vehicle blind spot detection
- Vehicle surrounding awareness
- New human-machine interface concept

However, these are kind of by-products since main intention was to develop common methodology for automotive software implementation. The project therefore, took steps forward in developing common framework (i.e. methodology) to bring new functions to the vehicles. These are not limited to above functionalities but they are the first steps.

The one DESERVE platform allows the co-design of software and hardware for applications and algorithms. The whole application or algorithm can be implemented in software using for example ADTF, RTMaps or Simulink interfaces which allows reusability, flexibility and fast verification of the implemented hardware modules.

1.5 Conclusions

The original project target was to develop a common software platform for modern vehicles. The expected outcome is that the platform fits up to 90 % of all new applications introduced in the new cars. The novel ADAS functions are becoming more and more complex and the new features are software-based instead of mechanical solutions like they were 10 to15 years ago. However, software is always prone to errors which may have serious consequences if e.g. the vehicle accelerates when emergency braking is expected. Therefore, a proper evaluation procedure is needed by using proper performance indicators, in order to verify the correct functionality of the platform.

As the final concluding remark, the DESERVE methodology pushes forward the situation compared to the current approaches in the automotive industry. The used architecture for the DESERVE platform is flexible and modular and enables to add new software components, devices, modules and functions even if the set of vehicle sensors, actuators and HMI remains.

PART I

ADAS Development Platform

2

The DESERVE Platform: A Flexible Development Framework to Seemlessly Support the ADAS Development Levels

Frank Badstübner[1], Ralf Ködel[1], Wilhelm Maurer[1], Martin Kunert[2], André Rolfsmeier[3], Joshué Pérez[4], Florian Giesemann[5], Guillermo Payá-Vayá[5], Holger Blume[5] and Gideon Reade[6]

[1]Infineon Technologies AG, Germany
[2]Robert Bosch GmbH, Germany
[3]dSpace GmbH, Germany
[4]INRIA, France
[5]IMS/Hannover University, Germany
[6]ASL, U.K.

2.1 Introduction to the DESERVE Platform Concept

As outlined by Figure 2.1, the DESERVE platform is the key enabler for speeding up the development of next generation ADAS systems. The DESERVE platform represents an open platform to be used by anyone. This chapter therefore covers the entire spectrum of aspects to be considered for the use of this generic DESERVE platform.

Please kindly note that the extensive work on the DESERVE platform cannot be completely described here. Thus, reference to a manifold of DESERVE deliverables are made. As most of these deliverables are not publicly available, essential findings in these deliverable reports were included here to provide a complete view on the DESERVE platform.

The DESERVE platform relies on model-based design and virtual testing tools. Its openness is based on the compliance with AUTOSAR standards. All AUTOSAR members have access to these standardized interfaces.

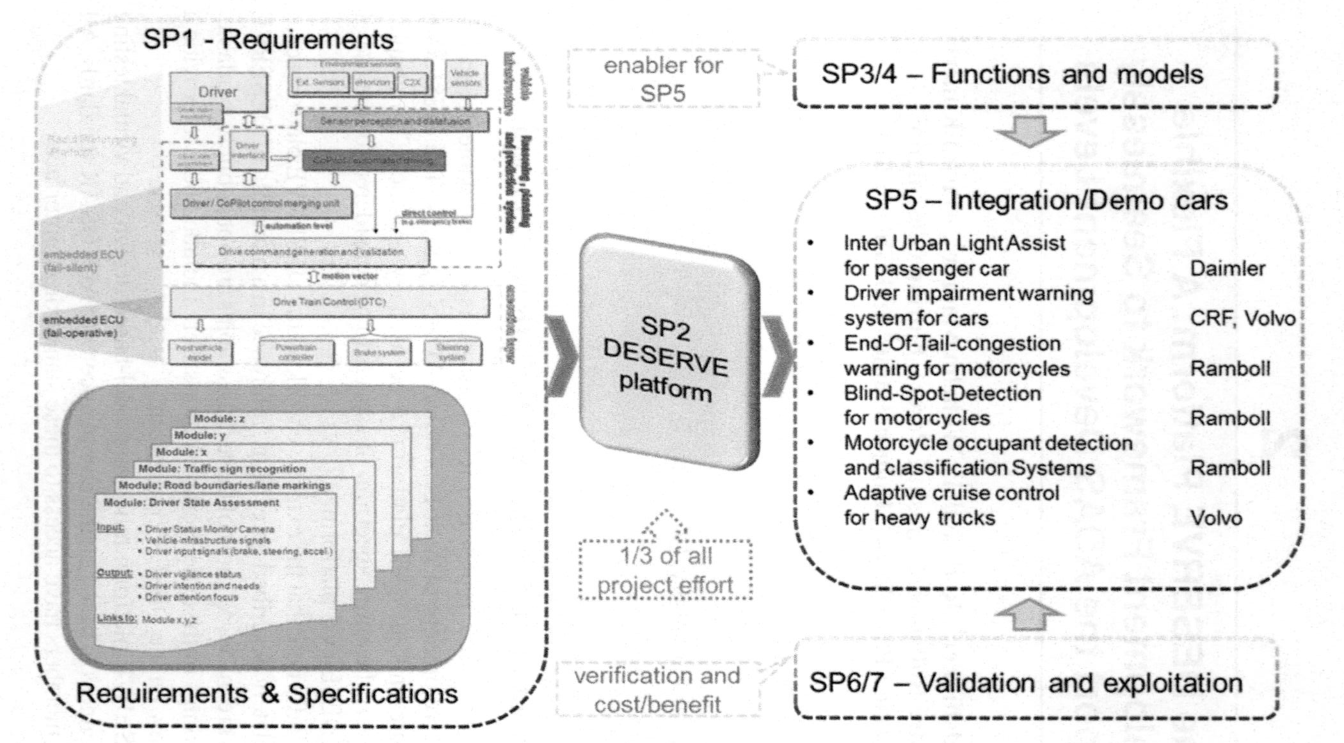

Figure 2.1 The DESERVE Platform – the enabler for next generation ADAS systems.

The DESERVE platform is not related to any specific hardware or software. In contrast, it is generic and represents a new methodology and concept to develop future ADAS systems more efficient and more flexible with maximum reuse of modules and components due to well-defined processes and standardizations on architecture and encapsulated module levels.

Requirements engineering is applied for next generation ADAS systems. By means of model-based design (e.g. Matlab/Simulink/ADTF/RTMaps) fast implementation in ADAS rapid prototyping framework is achieved (development level 2). Rapid prototyping results are evaluated by Hardware-in-the-Loop (HIL), Model-in-the-Loop (MIL) or Processor-in-the-Loop (PIL) test bench. In parallel, by making use of model based design space exploration, specifications and requirements for System-on-Chip (SoC) can be derived at a very early development phase, which supports cost prediction on basis of silicon area, throughput etc. Both, validation by virtual testing and cost prediction indicate important improvement potentials that need to be implemented in the next cycle of the iterative development process.

The situation before DESERVE can be characterized by the absence of model-based access to perception and fusion algorithms, missing AUTOSAR compatibility, there is no library with available algorithms (for composing and evaluating new algorithms). Rather, testing the application on real vehicles in real traffic scenarios is the approach followed, together with some recording feature to allow the capturing of the critical situations, where the solution fails for example, in order to reproduce them in some way later in laboratory.

The objectives of the DESERVE platform are driven by the market needs, which are enabling a further growth of embedded systems and more specifically advanced driver assistance systems (ADAS), mastering the complexity (both in system architecture and processing power) of ADAS, reducing costs of components and development time of ADAS as well as the seamless integration of the growing amount of functions within ADAS and the corresponding vehicle.

DESERVE strives to meet these markets needs by aiming at a novel design and more efficient development process that is enabled by a platform. A platform that provides a flexible development framework, reaching from early PC-based pre-developments down to close-to-production hardware implementations on final target systems on chip, to seamlessly support the ADAS development levels; that constructs a tool chain to allow for modelling and evaluation via virtual testing of new sensors, algorithms, applications and actuators during the whole design and development process; a platform; that forms a common in-vehicle platform for future ADAS functions based

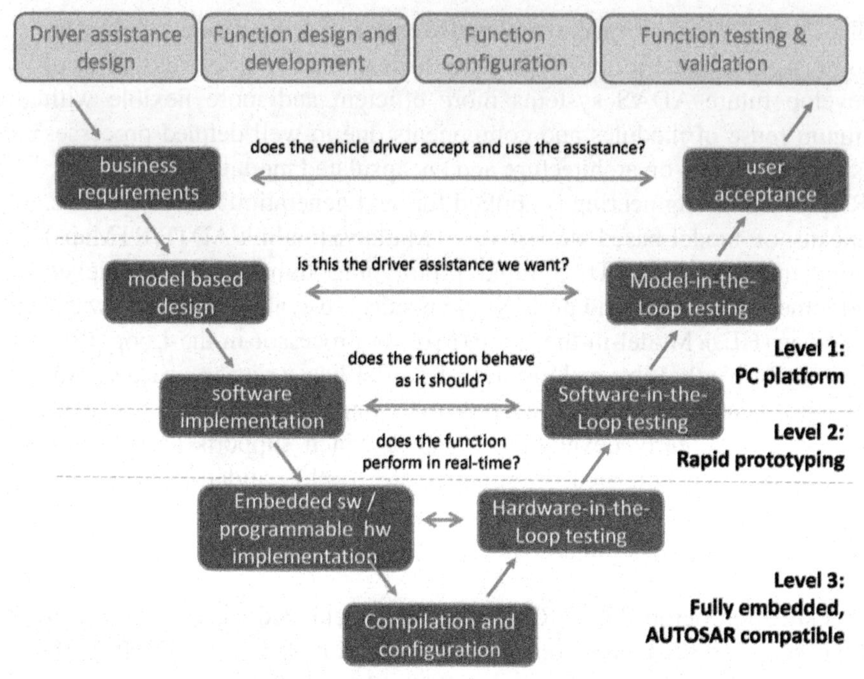

Figure 2.2 DESERVE platform enabled design and development process.

on a modular approach and an architecture and interface specifications that are compatible with AUTOSAR (access and easy-to-use also for non-project-partners); a platform that enables the integration of safety mechanisms for pre-certification (generic safety requirements e.g. for testing on public roads) and full requirements for ASIL D according to ISO 26262 (to prepare certification of later target platform) and security mechanisms for pre-certification of connected ADAS according to ISO 27001.

The novel design and efficient development process is based on the well-known V-model and fully DESERVE platform supported during all phases in the process. This is illustrated in Figure 2.2.

2.2 The DESERVE Platform – A Flexible Development Framework to Seamlessly Support the ADAS Development Levels

This section introduces into the development methods and guidelines associated with the DESERVE platform and outlines the benefits in terms of development cost and time savings from the OEM perspective. Basically, the platform concept is based on three pillars which reflect the different development levels and the transition of ADAS algorithms from the prototyping to production phase in the automotive industry (see Figure 2.3).

The DESERVE platform is a generic platform that supports all development levels illustrated in Figure 2.3 as seamless as possible – from feasibility study to product development.

Level 1: PC platform

In the research and pre-development phase users typically require highly flexible tools with an intuitive user interface and the implementation of ADAS algorithms may not satisfy hard real-time requirements. Here, PC-based tools such as ADTF and RTMaps for data fusion often constitute the basis for ADAS development.

Such tools provide a high user comfort and allow developers to implement and verify algorithms directly on a standard MS Windows or Linux PC. Different kinds of sensors/actuators and vehicle bus interfaces are available so that the algorithms can directly be tested in a real environment. However, real-time calculation is not guaranteed, especially with complex perception, fusion and tracking algorithms. In addition, there is no direct support of Matlab/Simulink, AUTOSAR and the model-based design approach for application functions. Finally, PC platforms as described above are typically not tailored for stand-alone, in-vehicle use cases.

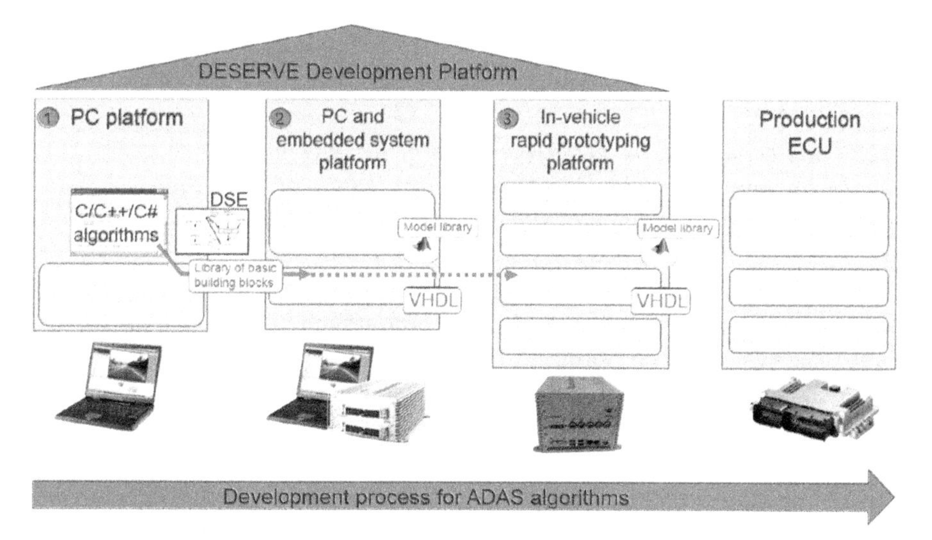

Figure 2.3 ADAS development process.

To avoid a time-consuming redesign of perception, fusion or tracking algorithms when implementing them on the final ECU hardware (production ECU), engineers are looking for ways to evaluate different target hardware architectures according to given cost criteria already in early development stages. This request is met by the design space exploration (DSE) methodology and the SoC modelling approach.

Level 2: Rapid prototyping platform including software superstructure (e.g. embedded PC/embedded controller with realtime operating system and FPGA)

In the second development stage engineers go one step closer to a real-time implementation. Complex and computationally intensive algorithms are shifted to a powerful FPGA to improve the realtime capability. In parallel to this, the FPGA platform allows different target hardware architectures to be evaluated in combination with the selected algorithms. To ensure a rapid implementation of the above mentioned perception, fusion, and tracking algorithms in the FPGA, basic building blocks in terms of a library are provided by the DSE framework. By means of this block-based modeling approach the time and effort for implementing the associated algorithms can significantly be reduced.

Using an embedded system platform in this stage featuring both an FPGA and an embedded controller also allows ADAS application algorithms to be designed by means of models so that the associated development time can further be reduced. Compared to the purely PC based framework real-time performance is almost guaranteed, though the user comfort with programming the FPGA may be restricted.

Level 3: Fully embedded, AUTOSAR compatible architecture (e.g. multicore controller with FPGA) for the evaluation of algorithms in realtime and implementation of safety requirements according to ISO 26262 (e.g. pre-certification for testing on public roads)

The goal of this stage is to go one step further to the final target hardware and to provide a stand-alone, in-vehicle rapid prototyping platform which, for example, can even be used during test drives. This stage reflects the users' need to evaluate and experience the driver assistance system directly in the vehicle itself.

The standard PC is replaced by an embedded PC that is qualified for in-vehicle use in terms of shock, vibration and temperature, similar to the other parts of the system. This platform also allows the integration of hardware accelerators so that even highly computational intensive algorithms may be tested in the vehicle. It is also possible to interface target microcontrollers of production ECUs and to run certain algorithms there. The complete platform behaves like a prototype ECU which can be operated by test drivers which are not specifically instructed. For example, the platform can be started and shut down via the vehicle's ignition key.

The development platforms of all stages can be used together with the model-based design space exploration approach for system on chip and libraries of basic building blocks for the FPGA. By means of this the gap is closed when transferring perception, fusion and tracking algorithms from prototyping to production, similar to the model-based design approach with application functions using Simulink. Being able to use already tested and validated building blocks and software modules greatly facilitates and expedites the development process.

To support the model-based development of algorithms at all processing layers (perception, decision making, warning and control strategies) and to execute these algorithms in the vehicle, the DESERVE platform level 3 needs to be fully compatible to the AUTOSAR standard (note: as of today, no certified AUTOSAR 4.0 real-time operating system including memory protection is available; its development is not subject of DESERVE).

In addition, at this development level, safety mechanisms need to be developed: According to ISO 26262 the DAS system needs to be classified concerning the Automotive Safety Integrity Level (ASIL). Many DAS systems require the highest classification ASIL D. Suitable measures are required to fulfil the related strong requirements. As the certification process is very much related to the hardware, just pre-certification (e.g. for testing of the new DAS on public roads) is possible at this development level.

As a result, OEMs are able to define early and precise enough the distinct requirements for the final ECU hard- and software (e.g. required interfaces – which I/O and bus system; computational power; memory requirements), including the safety mechanisms (e.g. memory protection, lockstep operation).

Level 4: Target production platform (e.g. multicore controller ECU with integrated custom ASIC/FPGA/hardware accelerator)

On basis of the production hardware, the final certification of the ADAS takes place. Within the DESERVE project, the generic DESERVE platform concept was validated. Starting with purely PC-based development, algorithms can be outsourced step by step to an FPGA or embedded controller prototyping system. In addition to the hardware concept, a design space exploration and an analytical modelling approach for system on chip is proposed. This software framework allows different target hardware architectures for the implementation of perception algorithms to be evaluated according to given cost criteria in early development phases. The software framework is coupled to the FPGA of the DESERVE platform. The associated workflow will be supported by a library of basic building blocks for the FPGA by means of which perception algorithms can be composed and implemented quickly.

To validate the platform concept, three different realization instances of the generic DESERVE platform are considered in the project:

- Level 1: Purely PC based solution
- Level 2: Mixed PC/embedded control based on dSpace Micro Autobox with FPGA framework (this platform will be extensively used for the ADAS vehicle demonstrators)
- Level 3: Fully embedded platform based on multicore controller plus FPGA. This instance of the DESERVE platform provides realtime operating system and basis software fully compatible to the AUTOSAR standard. Thus it is open and easy to use for all AUTOSAR members. It will also feature safety concepts required for ASIL D and consider new radar/camera interfaces.

2.3 DESERVE Platform Requirements

The next step in the definition process for the DESERVE platform concerned the translation of the previously defined platform needs into generic requirements for the DESERVE platform based on common software architecture and suitable for the development and simulation of the 33 DAS functions investigated in the beginning.

The generic requirements for the DESERVE platform were defined utilizing the following approach (see deliverables D1.2.1 [1]).

The DESERVE development platform has been defined taking into account that general requirements such as AUTOSAR compatibility [6], SPICE compliance and functional safety (ISO 26262) [7, 8] are mandatory for industrial use. These requirements apply for the "industrialized platform". The generic DESERVE platform addresses a functional software architecture based on Perception, Application and IWI platforms.

2.3.1 DESERVE Platform Framework

The DESERVE platform has been defined taking into account general requirements such as AUTOSAR compatibility, SPICE compliance and functional safety (ISO 26262), which are mandatory for the later industrial use. The AUTOSAR standard comprises a set of specifications describing software architecture components and defining their interfaces. DESERVE aims at using AUTOSAR to integrate applications from different suppliers inside a single processing unit.

DESERVE addressed also to be compliant with the SPICE standard, which represents a set of technical standards documents for the computer software development process and related business management functions. The ISO 26262 standard was considered in the implementation of DESERVE platform in order to improve the safety in the development of methods and tools. The ISO 26262 standard defines the “Functional Safety Assessment” at the completion of the item development with the scope to assess the functional safety that is achieved by the element under safety analysis.

The baseline for DESERVE is represented by the results of past and on-going research projects [9, 10], and in particular of interactIVe addressing the development of a common perception framework for multiple safety applications with unified output interface from the perception layer to the application layer [11].

Figure 2.4 presents the DESERVE platform framework. In this generic architecture the perception platform processes the data received from the sensors that are available on the ego vehicle and sends them to the application platform in order to develop control functions and to decide the actuation

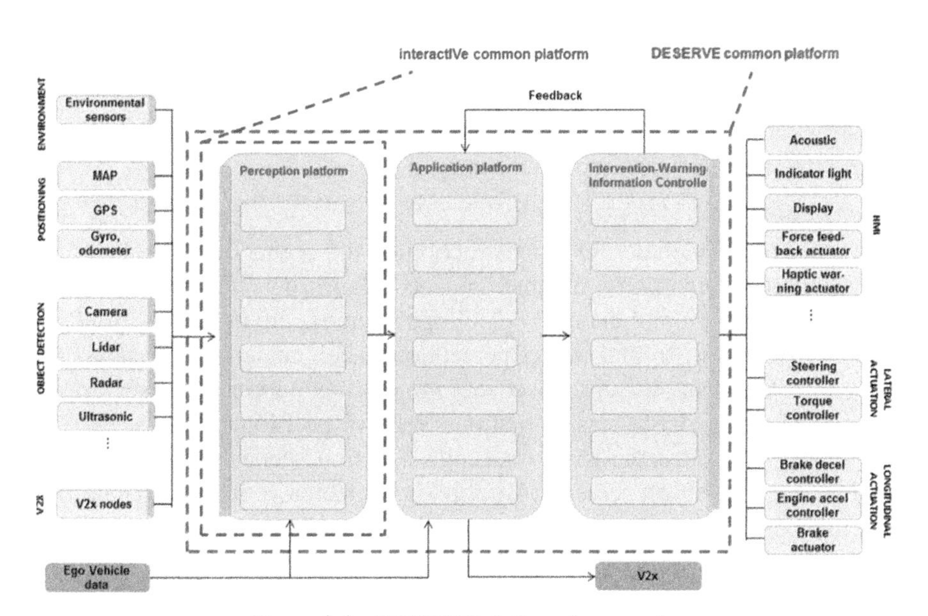

Figure 2.4 DESERVE platform framework.

strategies. Finally, the output is sent to the IWI platform informing the driver in case of warning conditions and activating the systems related to the longitudinal and/or lateral dynamics.

2.3.2 Generic DESERVE Platform Requirements (Relevant to all Development Levels)

Different clusters of requirements were defined following the structure of the DESERVE platform framework. Please note that each of the following requirements was divided in sub-requirements, which are described in detail in DESERVE deliverable D1.2.1.

General software requirements

General software requirements: Among others, these cover the previously mentioned software requirements for modularity, reusability, AUTOSAR, SPICE process assessment (ISO/IEC 15504), functional safety (ISO 26262), platform independence (the application software needs to be independent from the processing hardware), standardized interfaces (i.e. the software needs to have interfaces to sensors and actuators that are standardized and published), operating system independence (cross platform libraries are recommended), programming language, communication technologies independence, automatic start-up/shut-down, configuration of sensors position, software versioning and licenses.

General hardware platform requirements

These cover the aspects power supply, list of supported sensors, processing unit, unit size and number of included components etc.

Perception module requirements

These requirements include 3D reconstruction of the scene in front of the vehicle, ADASIS horizon, assignment of objects to lanes, detection of the free space, driver monitoring, enhanced vehicle positioning, environment, front near range perception, frontal object perception, lane course, lane recognition, moving object classification, occupant monitoring, parking lot detector, recognition of unavoidable crash situations, relative positioning of the ego vehicle to the road, road data fusion, road edge detection, scene labelling, self-calibration, side/rear object perception, traffic sign detector, vehicle filter/state, vehicle light detector, vehicle trajectory calculation, vulnerable road users detection and classification.

The functional architecture of the perception layer is illustrated in Figure 2.5. Depending on the ADAS system to be realized, some of the

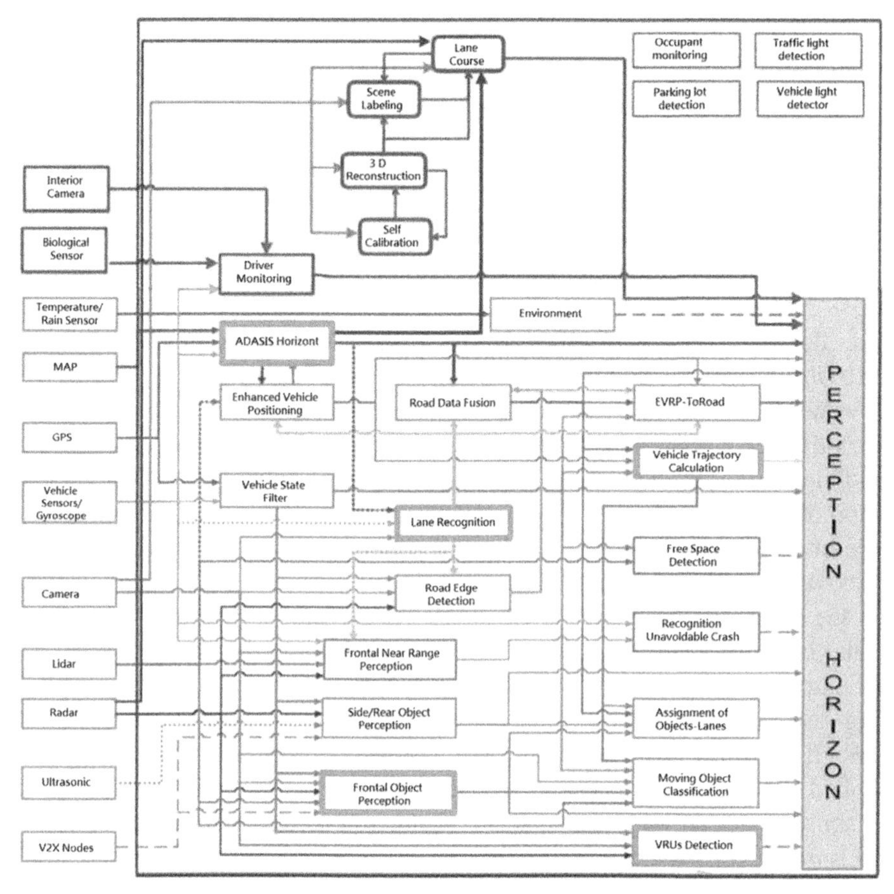

Figure 2.5 Perception platform functional architecture.

components in the generic perception platform architecture may be omitted (without losing generality). The modules developed in the project to build the demonstrators are highlighted by thicker boxes.

The number and variety of the different perception sources is manifold and requires special care and precaution to transport the available information in the subsequent data processing modules. Two main aspects have to be taken into consideration when connecting perception sources to the DESERVE platform: The information content may differ from sensor to sensor even when the same technique (e.g. radar, video camera or ultrasonic sensor) is used. Based on the physical concept used the individual sensors may have an intrinsic lack of information that can never be provided, independent of the effort spent to improve the sensor performance (e.g. radar sensors can never "visually" read the road signs content while video sensors can never provide direct speed measurements).

By using the general interface descriptor approach the data input structure for the perception layer processing module becomes independent from the real sensors connected to the DESERVE platform. This kind of concept is used in PC architecture since several years under the term hardware abstraction layer that completely decouples data information from the physical hardware in use.

The flexibility and scalability of the overall system is much better and reusability of SW components that are already developed is higher. Improvements and changes within the subgroups (i.e. environmental sensors or perception input processing module) can be conducted on a standalone basis without modifying or adapting the whole data processing chain at all. General adoption of the whole data processing chain is thus only needed in the case that the interference descriptors between the modules have to be updated or modified due to recently emerging needs.

As the diversity of the already existing environmental sensors is already huge and many products are already in series production, the change of the sensor output signals is often not possible at all. To connect already existing sensing devices or sensors with an IP-protected signal output to the open DESERVE platform, a work-around with converter or breakout boxes can be applied. Using such interface converter/breakout boxes almost any kind of sensor system can be attached to the standardized and abstracted input channels of the generic DESERVE platform.

Application module requirements

The application module needs to consider the following requirements: ACC control, activation control, advance warning generator, calculation of required evasion trajectory, decision unit, driver intention detection, driving strategy, intervention path determination, IWI manager, reference maneuver, situation analysis, target selection, threat assessment, trajectory control, trajectory planning, vehicle model and vehicle motion control.

The functional scheme of the application platform modules is depicted in Figure 2.6. The modules are divided in clusters having the same scope. Some of them have mainly the objective to select the driver intention and the most dangerous target. Other modules execute control operations and make an evaluation about the current situation of warning and eventually decide specific actions. Then the type of information to provide to the driver and the

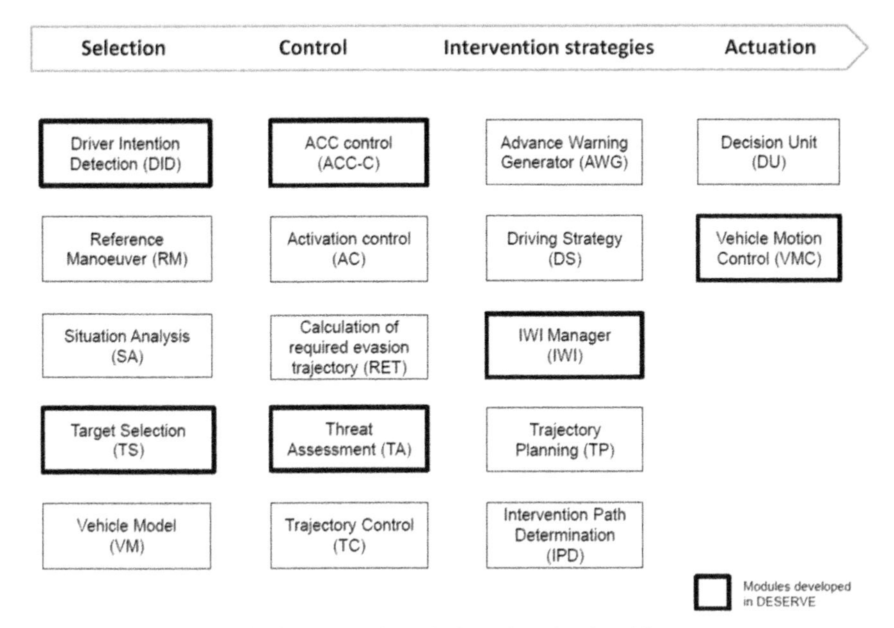

Figure 2.6 Application platform functional architecture.

intervention strategy are decided. Finally, the kind of actuation to adopt is provided to the IWI Platform modules.

IWI module requirements

The IWI module is dedicated to suit requirements regarding the HMI (acoustic, displays, telltales, haptic steering wheel, haptic accelerator pedal, haptic safety belt), actuation of external lights, lateral actuation (steering angle and steering torque controller) and longitudinal actuation (engine acceleration controller). The functional architecture of the IWI platform is depicted in Figure 2.7.

Different levels in the development process of ADAS require different instances (i.e. realizations) of the generic DESERVE platform – from PC based (development level 1) to production hardware (development level 4). With increasing development levels, additional requirements need to be addressed. This principle shall be explained in the next two subsections.

2.3.3 Rapid Prototyping Framework Requirements (Development Level 2)

This section shortly outlines the main requirements for the DESERVE rapid prototyping platform. The main intention here is to specify a flexible and

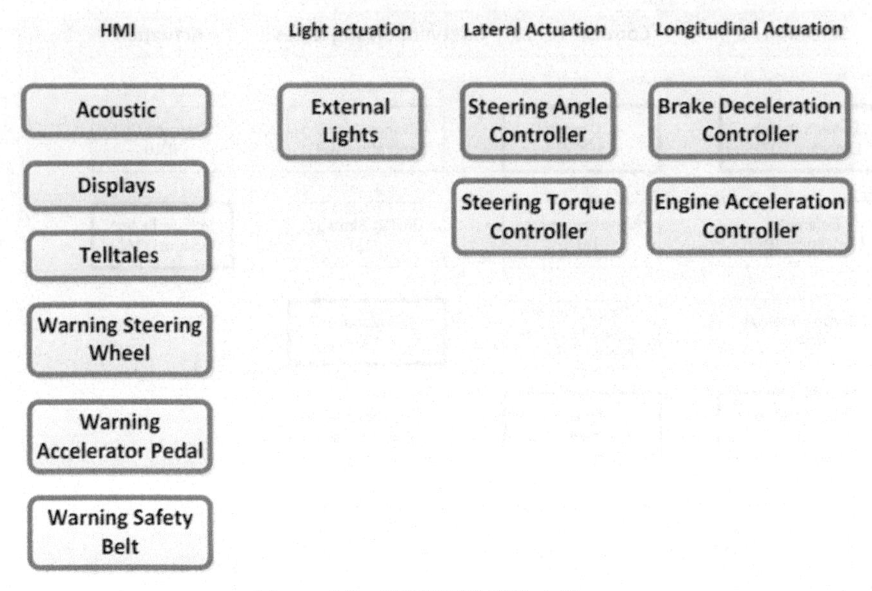

Figure 2.7 DESERVE IWI platform.

modular rapid prototyping environment allowing ADAS related perception, application and intervention algorithms to be developed in short iteration cycles and to be prototyped directly in the vehicle. In order to do so, there is a need to connect different kinds of sensors to the development framework, to pre-process and fuse the sensor data, to calculate the actual ADAS applications and to finally drive the respective actuators.

The structure for the generic requirements in the previous section, the rapid prototyping system requirements are structured in hardware, software and FPGA code requirements. In addition, a distinction is made between perception (i.e. sensor data processing) and application algorithms.

2.3.4 Additional Requirements for Embedded Multicore Platform with FPGA (Development Level 3)

While the main focus of development level 2 is on evaluation of algorithms in real-time on public roads, thus on ADAS functionalities and use in the DESERVE DAS function demonstrators, levels 3 (and 4) go significantly ahead in terms of fulfilling "critical" requirements like AUTOSAR compatibility, SPICE compliance and functional safety (ISO 26262) which are mandatory for industrial use of the platform. Due to limited resources and limited project duration, these requirements cannot be fully implemented in DESERVE. Nevertheless all the work done for the “non-industrialized” DESERVE platform can be (partly) reused or carried over to the industrialized version of the DESERVE platform (level 4).

2.4 DESERVE Platform Specification and Architecture

The generic platform requirements were translated into specifications, which represent the starting point for the development of modules for the DESERVE platform. The specifications were included into an Excel file which is accessible to all project partners via the project server. By means of an iterative process, both specifications and software design were refined and improved. A summary of the specification approach and of the specifications derived from the DESERVE platform requirements is provided in deliverable D1.3.1 [2].

2.4.1 DESERVE Platform Architecture

The architecture of the DESERVE development platform shall follow both the principle of standard DAS development cycles and the mappings of application building blocks to final, often heterogeneous hardware implementations. To date there is no tool or framework available that covers both requirements at the same time on the same platform.

In the early concept and implementation phase the basic development, specification and validation (e.g. with MIL, SIL or HIL) is often done with another development framework (both for SW and HW) than the one applied for the final target platform. Little is known or taken into account from the final embedded system characteristics when first application algorithms are programmed and very often the SW modules written in this first development environment have to be reprogrammed from the scratch when porting it to the embedded system on chip. If the software, mostly written in a high-level programming language, finally fits the target system one has selected for series production, is a game of pure chance and not rarely during the series product development cycle a larger target system or some “add-ons” have to be chosen. With the new design space exploration methodology the certainty to select the suitable embedded target system at first time is significantly increased.

The DESERVE development platform architecture has to comply with the following basic needs:

- Enough flexibility to encompass different development environments in a common, seamless framework for both the high-level algorithm development and the easy porting of these SW modules to the embedded target platform.

- Real time recording and playback capabilities for both the high-level and embedded system implementations.
- A communication architecture that is capable to shift SW portions from the high-level development side to the embedded target system as required (i.e. bypassing with HW accelerators).
- A seamless interoperability and replacement between the high-level (i.e. PC-based) and embedded target systems both for development and validation purposes.

The basic idea and intention of this hardware architecture is to standardize the interfaces between the three different development concept levels as good as possible.

Inputs from proprietary ADAS sensor systems and information sources are analyzed via a generic interface no. 1 to the PC based development environment. Here the ADTF tool with its filter programming concept is used to develop or improve SW modules on a high-level programming language. The partitioning and optimization of parts of the SW modules is consecutively done by shifting such portions over the generic interface no. 2 to the embedded controller framework that is already much nearer to the final commercial product. Via this bidirectional interface bypassing techniques like PIL (embedded Processor In the Loop) can be realized. In a final step, dedicated HW accelerators can be linked in via the generic interface no. 3 by applying the same bypassing concept. Especially computationally intensive tasks can so be "outsourced", so that even the PC-based platform is capable to keep the stringent real-time constraints.

Depending on the performance of the PC either all or only specific parts of the SW modules can be executed there. During the development process more and more SW parts are transferred to the HW-Accelerator level, which, in the final development stage, results in the next generation embedded ADAS target system. At this last development step, the level 1 (PC) and level 2 (embedded controller) platform will only serve as a shell to keep up the overall development framework.

Reuse of already existing components from former ADAS generations may be used in the early development phase as HW accelerators for computational intensive calculations. Mainly standard algorithms that are fixed and receive no further modifications are preferred candidates for such specific HW accelerators.

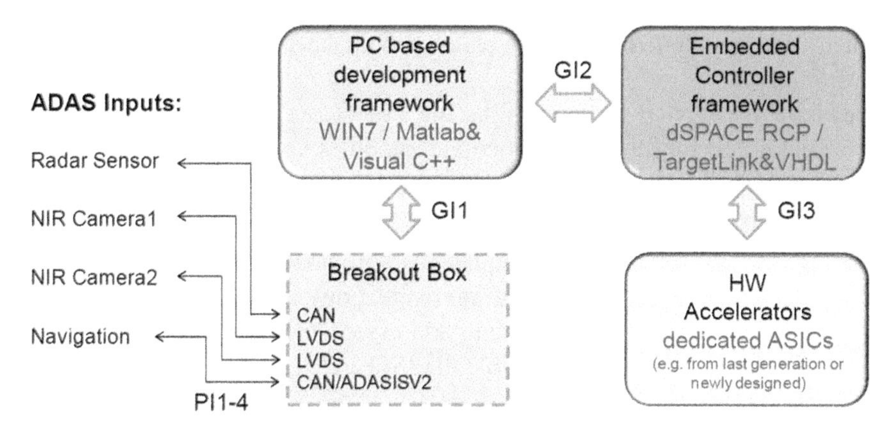

Figure 2.8 DESERVE platform (e.g. for development Level 2 – rapid prototyping system based on mixed PC and embedded controller framework).

This section summarizes the DESERVE platform architecture aspects. It considers hard- and software architecture aspects. The platform architecture is described in detail in deliverable D25.2 [4].

2.4.1.1 Hardware architecture

DESERVE has to be flexible enough to be implemented in a distributed and scalable architecture (several modules, each of them able to sense and/or process and/or actuate) or a concentrated one (sensors and actuators all linked with a single unit of processing and control). Task 2.5.1 identifies which conditions have to be satisfied by the individual subsystem architectures in order to be compliant with the DESERVE generic hardware platform.

For maximum reusability the DESERVE concept and hardware architecture was designed in such a way that subsystems of different generations (or respectively the kernels of it) can be used in parallel, thereby enabling the rapid and effective creation of next-generation innovative ADAS systems by using well tested and certified kernel functions of the "old" system which partly could be already implemented as SoC (System on Chip). The DESERVE development platform can be seen as a flexible rapid-prototyping environment that enables fast and efficient development of next generation ADAS functions in a continuous iteration cycle between the current and next-generation embedded subsystem components.

Furthermore, the DESERVE concept is flexible enough for different DESERVE partners to make different implementations. These would be of forms that might in future be interoperable, although DESERVE will not attempt to define detailed standards which would be necessary for actual interoperability.

The main DESERVE idea concerns the use of one common platform system (Figure 2.9) for all ADAS functional modules, instead of the current approach to have one platform for each individual ADAS system. Basically, three main hardware architecture challenges arise from this idea:

- Automotive quality: The platform needs to provide high reliability over the complete automotive temperature range, power supply and environmental conditions. As ADAS systems address safety aspects, the platform should implement as far as possible the ISO 26262 requirements, i.e. at least the hardware components that are near to the final product unit shall support the required ASIL level.
- Possibility to extend hardware capabilities: The platform needs to be designed up-front to support the possibility to include additional hardware into the system. Standard sensor interfaces are needed, for instance, but also standardized interfacing to external FPGA/DSP for performance enhancement is required. For scalability purposes, such external devices need to be cascadable. Similar considerations hold for the memory interface capability.
- A special case of hardware extension capabilities is the reuse of serial parts from earlier generations to speed up the development process or to increase the sensor perception by placing more sensors on the car.
- Finally, a seamless environment tool chain is needed. One key requirement lies in the reuse of the existing tool ecosystem over several platform generations. Further, we should target adaptability of the tools to the broad industry use cases, e.g. next generation video and radar sensors. Additionally, real-time monitoring and debugging of interface and processing for development purposes represent key challenges.

2.4.1.2 Software architecture

As for hardware architecture, the characteristics and constraints that the software architecture has to fulfill to accept an application based on modules developed inside the DESERVE platform (Figure 2.10) were identified. AUTOSAR standards were considered[1].

[1]Note: Being a research project, the development work conducted in DESERVE is discharged from being fully compliant with the AUTOSAR standard. Where possible and easy to implement, inputs from AUTOSAR were considered, of course. A mandatory request for AUTOSAR compliance is, however, not up for discussion.

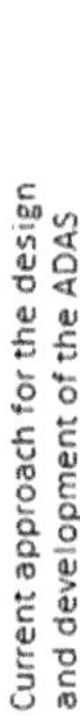

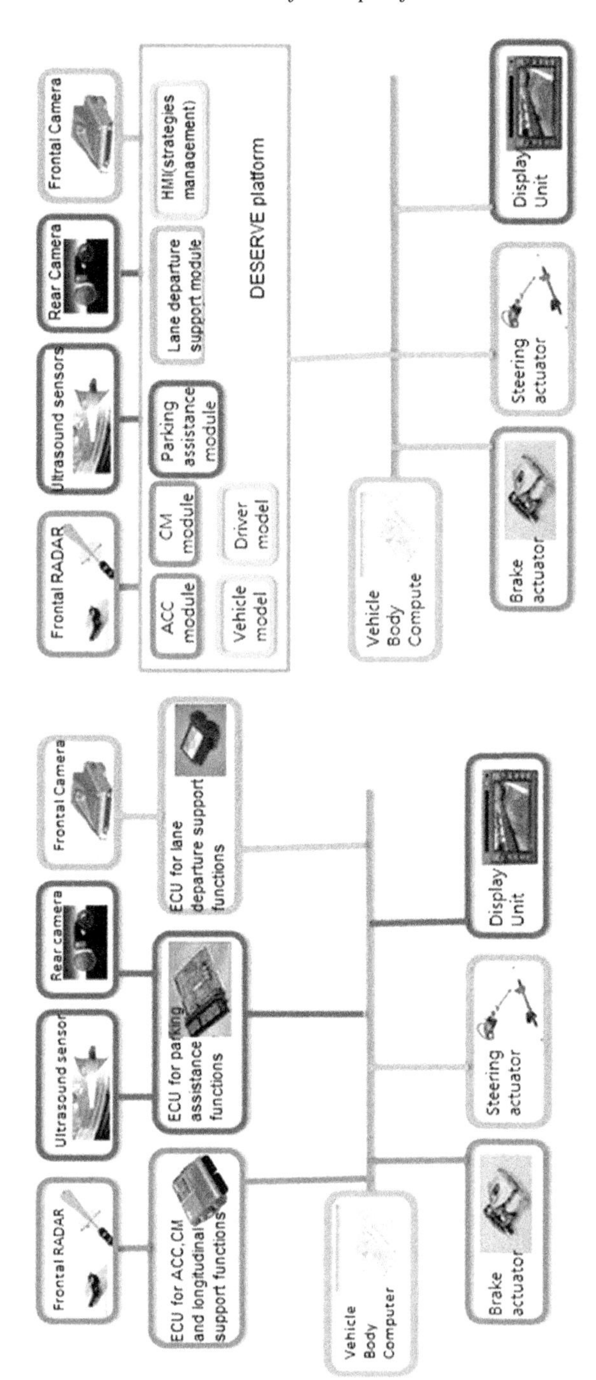

Figure 2.9 DESERVE approach – use of common platform for all ADAS modules.

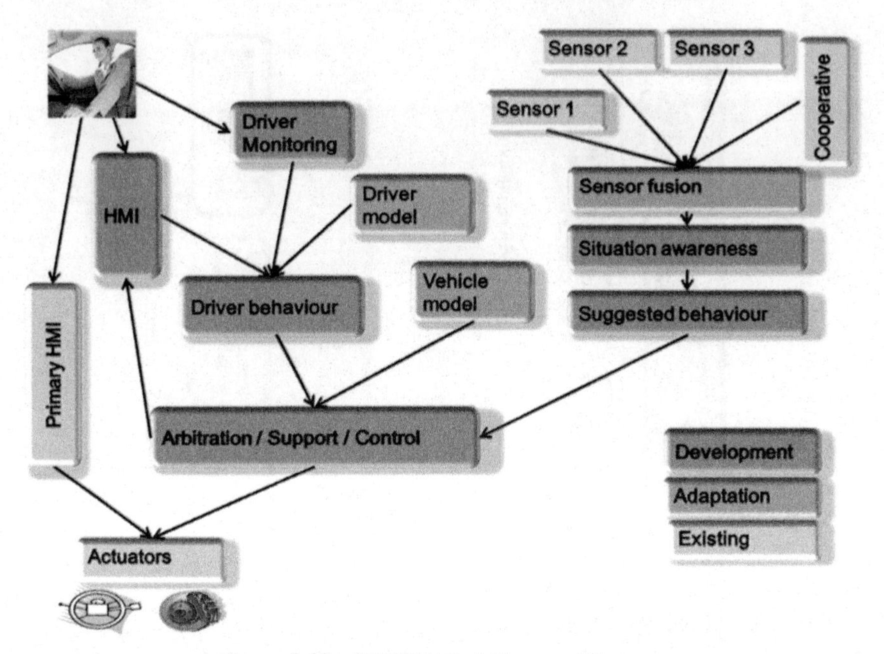

Figure 2.10 DESERVE platform architecture.

The key architecture challenges are: AUTOSAR Standards Architecture for the full platform system including performance accelerators, request for high SW re-usability/testability including re-use of older generation software blocks, fast time to market, highly optimized library for optimal performance, automatic code generation, standard compiler/tool chain and finally, hardware tool software support for realtime debugging, high speed parallel sensor data capture for validation and on-system debugging is required.

Application Software Modules

On the base of AUTOSAR standard, the general software architecture can be represented in three main layers: low level (basic software: this level abstracts from the hardware, provides basic and complex drivers and services for high level, i.e. memory, I/O), middle level (virtual function bus and runtime infrastructure) and high level (application software components).

The AUTOSAR standard introduces two architectural concepts (respects to other embedded software architectures) that facilitate infrastructure independent software development. Namely, these are the Virtual Function Bus (VFB) and the Runtime Infrastructure (RTE) that are closely related to each other.

In order to realize this degree of flexibility against the underlying infrastructure, the AUTOSAR software architecture follows several abstraction principles. In general, any piece of software within an AUTOSAR infrastructure can be seen as an independent component while each AUTOSAR application is a set of inter-connected AUTOSAR components.

Further, the different layers of abstraction allow the application designer to disregard several aspects of the physical system on which the application will later be deployed on, like type of micro controller, type of ECU hardware, physical location of interconnected components, networking technology/buses or instantiation of components/number of instances.

The middle level, VFB (Figure 2.11), provides generic communication services that can be consumed by any existing AUTOSAR software component. Although any of these services are virtual. They will in a later development phase be mapped to actual implemented methods that are specific for the underlying hardware infrastructure. The RTE (runtime environment) provides an actual representation of the virtual concepts of the VFB for one specific ECU.

An AUTOSAR software component in general is the core of any AUTOSAR application. It is built as a hierarchical composition of atomic software components. The AUTOSAR software component can be divided in Application Software Component and AUTOSAR Interface. It is important for DESERVE to preserve (and build up during the prototyping phase of the applications) the AUTOSAR modularity concept. Consequently, DESERVE focuses on the development of modular Application Software Components.

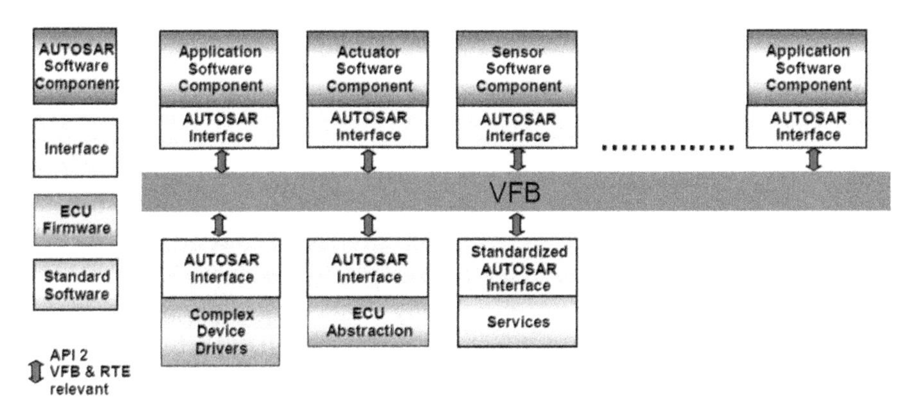

Figure 2.11 Overview on the principles of virtual interaction using the AUTOSAR.

Multi-task option to permit adding and removing of functionalities
The modularity is one the most important directive in the design of a global architecture, their functions and modules for embedded systems. Different multi-tasks (called processes) can be executed by sharing common processing resources in the same CPU. In this line, multi-thread languages as C++ are used by different developers around the world.

The software environments used in the DESERVE platforms (e.g. ADTF and RTMaps) are able to transfer functions already programmed in C and C++. These tools are multi-sensory software, designed for fast and robust implementation in multitask systems. They use functional blocks (called components) for data flowing between different types of modules: video, audio, byte streams, CAN frames, among others.

This multi-threaded architecture allows the use of multiple asynchronous sensors within the same application (see RTMaps and ADTF sections in D1.3.2 [3]). Moreover, they take advantage of multi-processor architecture for more computing power.

Based on the Development Platform Requirements [1], there are three main stages in the control architecture: perception, application and IWI platform. The goal of the DESERVE approach is to add different functions (Multi-task) in the same platform.

2.4.2 DESERVE Platform Interface Definition

The definition of the DESERVE interface architecture is described together with state of the art ADAS interfaces and next generation interfaces in deliverable D2.5.4 [5]. Due to the high relevance of the interface architecture for the DESERVE platform concept, a brief description is included in the next paragraphs.

2.4.2.1 Definition of DESERVE interface architecture

The definitions of the interface architecture plays a central role for the communication and data exchange between the different DESERVE platform modules and sensor components. In the DESERVE deliverable D2.2.1 [12] the abstracted interface descriptors are already defined on a content-based hierarchical level. With standardized information data flow between the numerous platform modules both the development time and the extension in performance and scope of the encapsulated modules can be realized very efficiently and in a well-structured way. The architecture of the interface has to be defined individually for each of the existing OSI layers, starting from the physical layer up to the application layer.

For modules that only communicate within the same hardware unit the physical data and communication layer are no longer needed. Instead, a message box oriented data transfer link is proposed for usage in the DESERVE project. The data to be transmitted is written in a predefined message box descriptor field and message flags trigger the synchronization and data updates in the concerned modules. The message box principle is sketched in Figure 2.12.

The interfacing concept of the AUTOSAR standard is considered and incorporated in the DESERVE platform where useful and appropriate. The AUTOSAR mode of operation, as depicted in Figure 2.13, fits already quite well with the general DESERVE approach proposed in this document.

In order to achieve a good reusability of embedded software functions, it has proven to be efficient in the industry to separate the "function software" from parameters defining the behavior of the software (= calibration data). This allows generating embedded systems with generic software functionalities by "embedded systems suppliers" (e.g. Continental, Bosch or others). Such systems are bought by OEMs for building their ADAS systems. The OEM can adapt the generic function to the individual behavior significant for his customers "just by calibration". In this process via an application system (market leader is INCA for example), the calibration data can be changed while the embedded system is running – regardless if simulated on a PC or

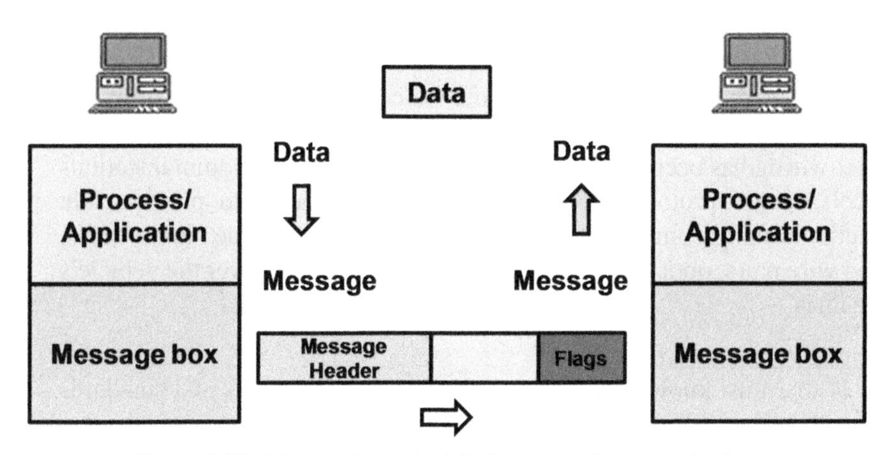

Figure 2.12 Message box principle for intra-unit communication.

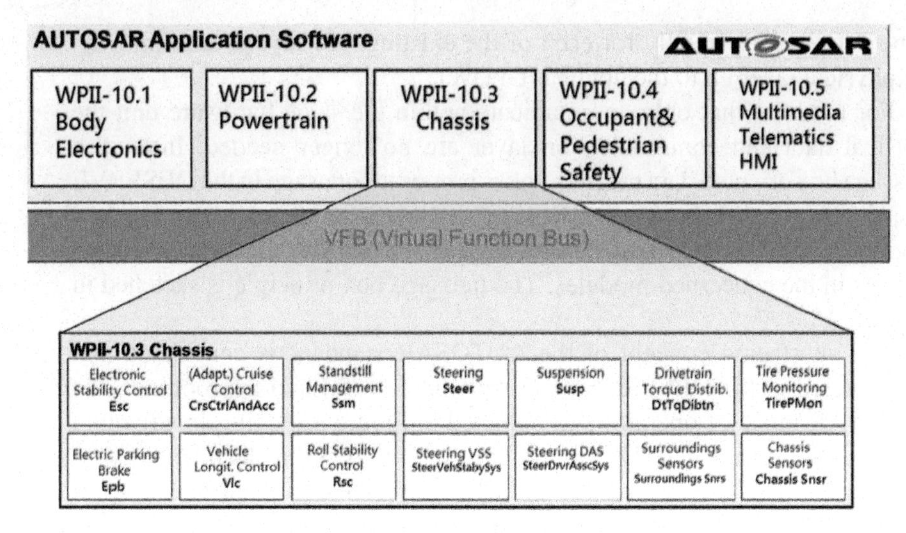

Figure 2.13 AUTOSAR application software concept.

running already on the target hardware. The separation of calibration date and function software is also allowed according to the AUTOSAR concept.

2.4.2.2 Existing ADAS interfaces

All electronic embedded systems used to control vehicle functions (specifically ADAS) need communications networks and protocols to manage all the process information. The modules receive input information from a network of sensors (e.g. for engine speed, lasers, cameras, etc.) and send commands to the control stage (Application platform in DESERVE), and finally to the actuators or warning systems that execute the commands (IWI platform) [1].

Due to the increasing complexity of modern ADAS applications, point-to-point wiring has been replaced by multiple networks and communications protocols. These protocols use different physical media to provide safe connection among components on the vehicle. These include single wires, twisted wire pairs, optical fiber cables, and communication over the vehicle's power lines.

Communication protocols

Some of the most known and used communication protocols and standards used in nowadays vehicles are:

- CAN (controller area network)
- VAN (vehicle area network)

- FlexRay
- LIN (local interconnect network)
- SAE-J1939 and ISO 11783
- MOST (Media-Oriented Systems Transport)
- Keyword Protocol 2000 (KWP2000)

Recent vehicles have installed multiple networks (with different protocols) to communicate among electronic control units (ECU) onboard. The networks are isolated from one another for several reasons, including bandwidth and integration concerns.

Existing interface standards

Current ADAS systems are designed and built to provide a dedicated answer to specific functionalities. Most ADAS are including in the same box the sensor itself and the processing unit. So, the raw data provided by the sensor (camera, radar) are directly loaded inside the ECU unit and processed. Only high level (processed) information is available on the communication buses. Raw data (e.g. pixel information of images) is not available.

The ADAS modules are dedicated products which communicate mainly within the same hardware unit. Nevertheless, to adjust the algorithms in function of the vehicle status, it's necessary to provide the ADAS modules with some vehicle information as: speed, yaw rate, direction indicator status, etc.

To manage the vehicle information acquisition and sending of the outputs, various communication interfaces are available, depending on the product, e.g. CAN or FlexRay communication interfaces.

The communication bandwidth requirements increase more and more with more and more complex applications, the existing network are not specified to cover the increasing demands for bandwidth, and the Ethernet price. Ethernet seems to be an alternative to the existing communication hardware.

2.4.2.3 Definition of next generation interfaces

The definition of next generation high speed sensor interfaces is the key to enable the improvement for next generation driver assistant systems. An optimized interface leads to optimized dataflow and system performance. For each sensor family (Camera/RADAR) there is a dedicated interfacing needed.

Parallel camera interface (CIF)

The Camera Interface (CIF) represents a complete video and still picture input interface transferring data from an image sensor into video memory. Furthermore, several hardware blocks – performing image processing operations on the incoming data – are provided (Figure 2.14).

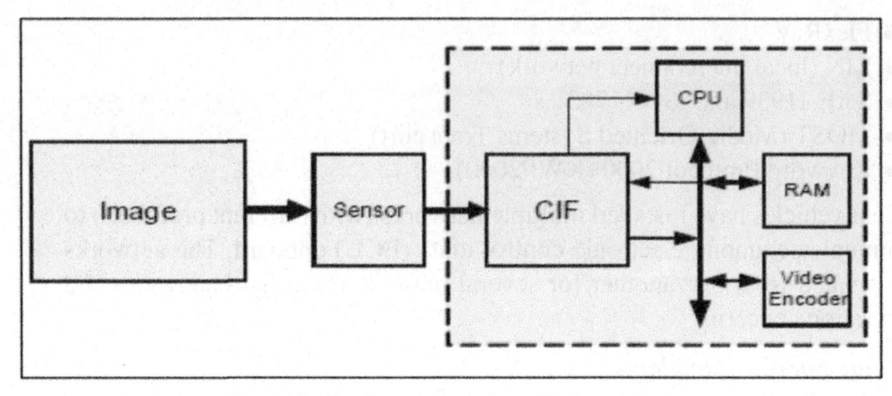

Figure 2.14 Camera Interface (CIF) overview.

Apart from providing the physical interfacing to various types of camera sensor modules, the CIF block implements image processing and encoding functionalities. The integrated image processing unit supports image sensors with integrated YCbCr processing. Additionally, the CIF also supports the transfer of RAW (e.g. Bayer Pattern) images and non-frame synchronized data packets. The CIF block features a 16 bit parallel interface. All output data are transmitted via the memory interface to a BBB (Back Bone Bus) system using the master interface. Programming of the CIF is done by register read/write transactions using a BBB slave interface.

The CIF provides a sensor/camera interface for a wide variety of video applications and it is optimized for high speed data transmission under terms of low power consumption. This module is designed to be used for the following use cases: video capturing/encoding, still image capturing in YCbCr with on-the-fly JPEG encoding and RAW frame data capturing.

The CIF requires fast system memory for image storage in either planar, semi-planar or interleaved YCbCr or RAW planar format or as JPEG compressed data. The iJPEG encoding engine should be able to generate a full JFIF 1.02 compliant JPEG file that can be displayed directly by any image viewer. Important YCbCr formats – which are used for video compression (e.g. MPEG4) for instance – are supported. For on-the-fly encoding macro block line interrupts are generated to trigger video encoding.

Serial RADAR interface (RIF)

Analog-to-digital converter (ADC) sample rates have been increasing steadily for years to accommodate newer bandwidth-hungry applications in communication, instrumentation, and consumer markets. Coupled with the need to digitize signals early in the signal chain to take advantage of digital signal processing techniques, this has motivated the development of high-speed ADC cores that can digitize at clock rates higher than 100 MHz to 200 MHz with 8 to 12 bit resolution.

In standalone converters, the ADC needs to be able to drive receiving logic and accompanying PCB trace capacitance. Current switching transients due to driving the load can couple back to the ADC analog front end, adversely affecting performance. One approach to minimize this effect has been to provide the output data at one-half the clock rate by multiplexing two output ports, reducing required edge rates, and increasing available settling time between switching instants.

Use of LVDS for ADC high speed data output

A new approach to providing high-speed data outputs while minimizing performance limitations in ADC applications is the use of LVDS (low voltage differential signaling). Infineon is incorporating LVDS output capability in new RF devices ADCs—and will include LVDS input capability in its new micro-controller designs.

Standards

Two standards have been written to define LVDS. One is the ANSI/TIA/EIA-644 which is titled "Electrical Characteristics of Low Voltage Differential Signaling (LVDS) Interface Circuits." The other is IEEE Standard 1596.3 which is titled "IEEE Standard for Low-Voltage Differential Signals (LVDS) for Scalable Coherent Interface" (SCI).

Generic interface to communicate between ADTF project and FPGA based hardware platform

In order to allow an easy and standard communication between an ADTF-Project and the FPGA-based hardware platform, a generic interface is used. The generic interface realizes the communication with different processing elements implemented in the FPGA-based hardware platform transparent to the user.

2.5 Safety Standards and Certification Concepts

Some concepts related to modular certification have already been adopted by current standards and thus have found their way into the state of the practice. This is particularly true for the fields of automotive systems because the trend towards modularized architectures has been particularly strong in this field.

2.5.1 Safety Impact of DESERVE

Modularization of a common ADAS platform comes with a clear impact on safety. Modules will interact, for example on Missed Trigger Interaction, Shared Trigger Interaction, Sequential Action Interaction and/or Looping Interaction.

Module interaction implies that any change in operation of one module (feature) can be attributed in part or in whole to the presence of any other module (feature) in the operational environment, as illustrated in the Figure 2.15.

2.5.2 Functional Safety of Road Vehicles (ISO 26262)

The international standard ISO 26262 for the functional safety of street vehicles contains the so-called concept of Safety Element out of Context (SEooC).

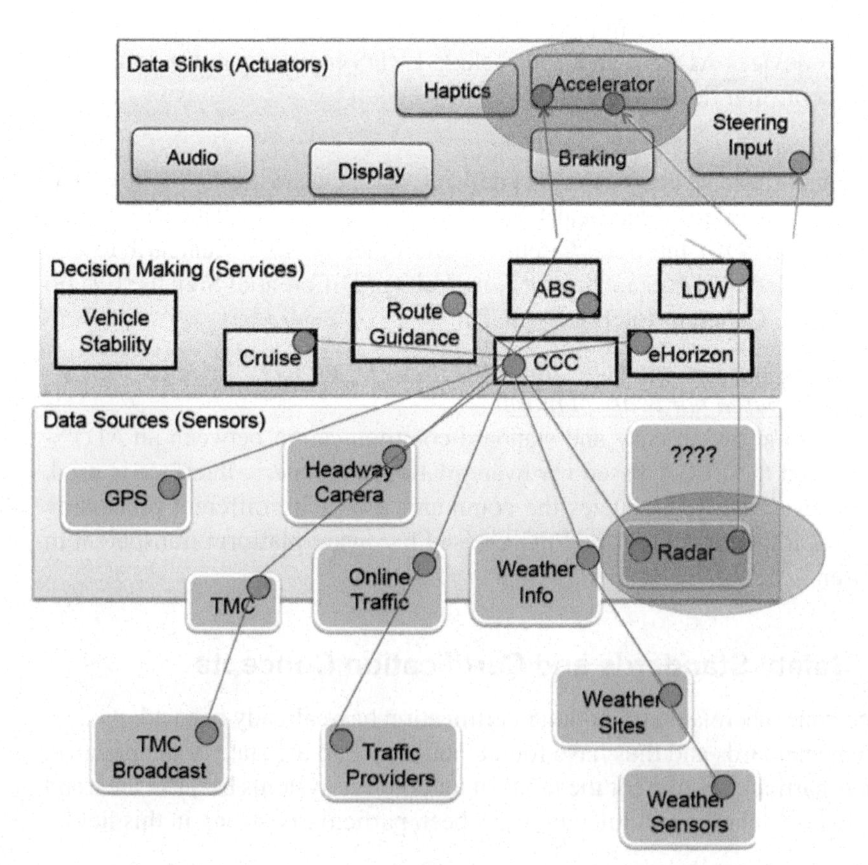

Figure 2.15 Module interaction implies changes in system behavior.

A SEooC is defined as a component for which there is no single predestinated application in a specific system. Therefore, the SEooC developer does not know the concrete role the product has to play in the safety concept. Subsystems, hardware components, and software components may be developed as SEooCs. Typical software SEooCs are reusable, application independent components such as operating systems, libraries, or middleware in general.

For SEooC development, the standard suggests specifying assumed safety requirements and developing the system according to these requirements. When the SEooC is to be used in a specific system, the system developer has to specify the demanded requirements, which can subsequently be checked against the assumed requirements. If there is a match between the demanded and the guaranteed (assumed) requirements, system and component are compatible.

The standard does not provide any suggestions or methods on how to identify safety requirements such as to increase the chance that assumed and real requirements will actually match. The standard specifies a relatively coarse-grained process for embedding a SEooC development into the standard's safety lifecycle. This approach deals with hierarchical modularization since it focuses on the SEooC's role as a sub-component of a system.

In general, integration of the SEooC is expected to be done at development time and thus there is no explicit support for open systems where components are to be integrated dynamically.

2.5.3 Guidelines Related to ISO 26262

ISO 26262 is a derivative of IEC 61508, the generic functional safety standard for electrical and electronic (E/E) systems. Ten volumes make up ISO 26262. It is designed for series production cars, and contains sections specific for management, concept and development phase, production, operation, service and decommission.

The ISO 26262 requires the application of a "functional safety approach", starting from the preliminary vehicle development phases and continuing throughout the whole product lifecycle.

The DESERVE project focuses on the concept and development (at system, hardware and software level) phases of the lifecycle. During these phases, the main steps defined by the Standard are:

Item definition: the Item has to be identified and described. To have a satisfactory understanding of the item, it is necessary to know about its functionality, interfaces, and any relevant environmental conditions.

Hazard analysis and risk assessment: to evaluate the risk associated with the item under safety analysis, a risk assessment is required. The risk assessment considers the functionality of the item and a relevant set of scenarios. This step produces the ASIL (Automotive Safety Integrity Level) level and the top level safety requirements.

The ASIL is one of the key concepts in the ISO 26262. The intended functions of the system are analyzed with respect to possible hazards. The ASIL asks the question: "If a failure arises, what will happen to the driver and to associated road users?".

The risk of each hazardous event is evaluated on the basis of frequency of the situation (or "exposure"), impact of possible damage (or "severity") and controllability.

The ASIL level is standardized in the scale: QM: quality management, no-risk and A, B, C, D: increasing risk with D being the most demanding. The ASIL shall be determined without taking into account the technologies used in the system. It is purely based on the harm to the driver and to the other road users.

Identification of technical safety requirements: the top level safety requirements are detailed and allocated to system components.

Identification of Software and Hardware safety requirements: The technical safety requirements are divided into hardware and software safety requirements. The specification of the software safety requirements considers constraints of the hardware and the impact of these constraints on the software.

To take into account the functional safety approach, the DESERVE applications should consider the application of the following main points: analyze risk early in the development process; establish the appropriate safety requirements and consider these requirements in software and hardware development.

The impact of the standard is different for the development of warning functions, control functions or automated driving functions.

2.5.4 Safety and AUTOSAR

In the automotive domain, Östberg and Bengtsson [14] propose an extension to AUTomotive Open System Architecture (AUTOSAR) which consists of a safety manager that actively enforces the safety rules described in dynamic safety contracts. Their main contribution is a conceptual model of safety architecture suitable for runtime based safety assessment. Openness and Adaptivity were both addressed.

Also in the automotive domain, Frtunikj et al. [15] present a runtime qualitative safety assessment that considers Automotive Safety Integrity Level (ASIL) and its decompositions in open automotive systems. In their solution, the authors consider the modularization of safety-assessment using Safety Elements out of Context (SEooC) from ISO 26262. In their approach, the SEooC was extended and the safety-assessment is done at runtime by a Safety Manager component.

2.5.5 Safety Mechanisms for DESERVE Platform

As an example, this paragraph summarizes some features of the safety mechanisms that are available by Infineon's multi-core platform AURIX which represents a potential instance of DESERVE platform (development level 3). Its safety documentation includes:

- Safety case report providing the arguments with evidence that the objectives of the ISO 26262 and the safety requirements for a component are complete and satisfactory.
- FMEDA (customer and Infineon proprietary document)
- Safety manual including an overview of the assumed application use cases and guidance for the application level, a summary of safety features and mechanisms and their recommended use as well as the summary of achieved safety metrics and resulting ASIL compliance [13].

The AURIX microcontroller platform is developed as a SEooC (Safety Element out of Context) and provides the safety mechanisms summarized in Figure 2.16. It provides a Safe Computation Backbone compliant with ISO 26262 ASIL D (this includes Single Point Fault Metric fully supported by HW mechanisms and Latent Fault Metric supported by SW (SafeTlib), Logic MIST, MBIST). Support criteria for coexistence of elements are enabled through a layered protection system (covering CPU tasks, Shared Memories, Peripherals), CPU supervisor/user privileges, Safety Task Attribute and a rich set of counters & watchdogs for program flow & temporal monitoring. SEooC deliverables are the Safety Library (SafeTlib), Safety Manual to support SEooC integration and FMEDA to support computation of the ISO 26262 Metrics.

Top Level Safety Requirements (TLSR) related to the Microcontroller I/O sub-system are specified by the system integrator, as these vary for

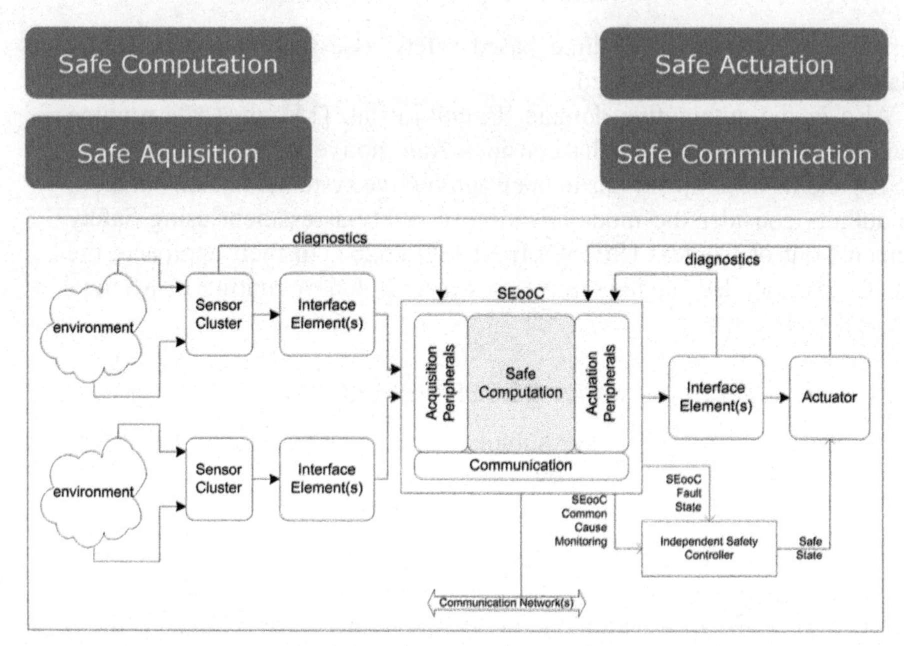

Figure 2.16 SEooC safety mechanisms.

each application. TLSR1 (ASIL D) requires to avoid false output of the microcontroller for longer than the FTTI (Fault Tolerance Time Interval, Figure 2.17), while TLSR2 (ASIL B) only require to avoid unavailability of a safety mechanism for longer than one driving cycle.

The Fault Tolerant Time Interval is more precisely defined by Figure 2.18. The application dependent fault detection time worst case is the diagnostic time interval. The fault detection time depends on the safety mechanism. The fault reaction time is the sum of failure signaling time and failure reaction time. Failure signaling time depends on the microcontroller architecture, while failure reaction time depends on the application. The failure signaling time is composed by the alarm forwarding time plus the alarm processing time plus the failure signaling time.

Safety requirements

With the AURIX as basis for DESERVE platform realization, it fulfils the targets according to ISO 26262-5, 8.4.5, which defines requirements for ISO 26262 metrics. To achieve ASIL D, for instance, the single point failure metric (SPFM) needs to reach minimum 99% and the latent fault metric (LFM) needs to reach 90% or above. The minimum values of SPFM and LFM shall

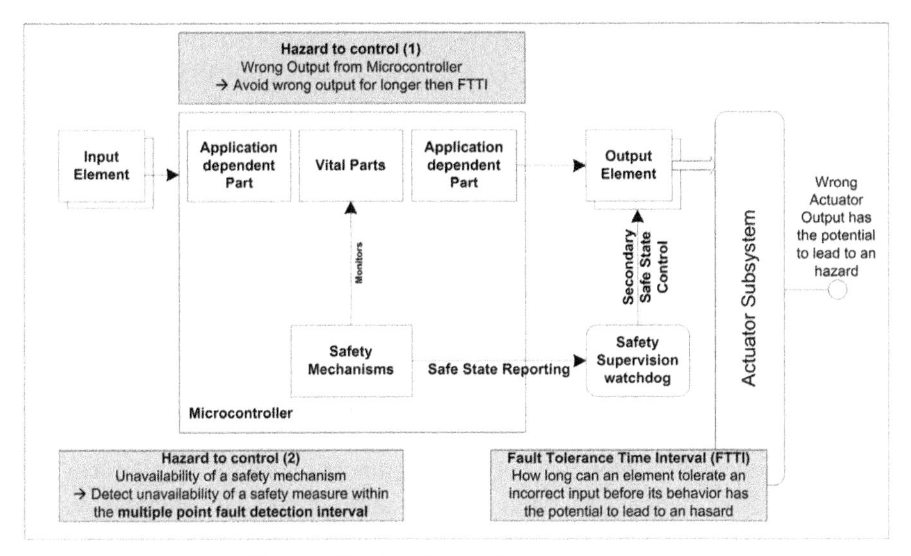

Figure 2.17 Top level safety requirements.

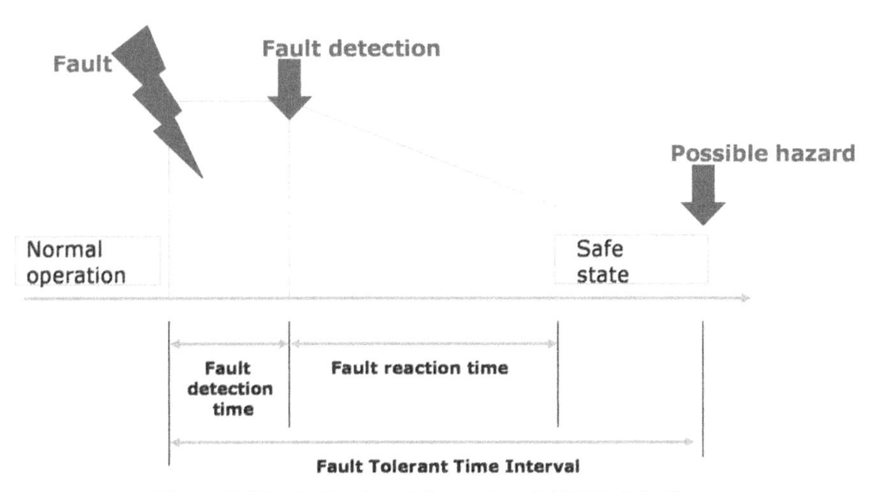

Figure 2.18 Fault tolerant time interval (FTTI) definition.

be reached by every vital part. The SPFM threshold levels shall be reached both for permanent and for transient faults. For a given ADAS application SPFM, LFM and PMHF (probabilistic metric related to hardware failures) metrics are estimated based on the vital, critical and application-dependent parts utilization.

In terms of PMHF for ASIL D safety goal, ISO 26262-5 requires a metric of less than 10 FIT (failure in time, referring to 10^9 hours). ISO 26262-5 9.4.3.6 and 9.4.3.7 specify the relationship between ASIL and FCR and DC (Residual Faults). To meet ASIL D requirements the diagnostic coverage for a FCR5 part shall be > 99.99%. The safety mechanisms are designed to achieve coverage of 99.99%.

Safety architecture

The safety architecture goal is to provide a safe computation platform for up to ASIL D safety applications according to ISO 26262, as this ASIL level is required for most next generation ADAS. To achieve this level, safe computation hardware and software, safe operating system as well as safe software architectures are required.

The generic elements (vital parts) of a safe computation hardware platform are summarized in Figure 2.19. Safe CPU requires hardware redundancy, realized by delayed lockstep CPU with enhanced timing and design diversity. Safe SRAMs allows information redundancy (realized by standard SECDED ECC, address signatures). Also safe Flash memory is needed for information redundancy (realized by an enhanced ECC with more than 99% coverage of arbitrary multiple-bit fault). Enhanced error detection codes for covering data & addressing faults lead to safe interconnects and support information redundancy. The clock system frequency range monitors using internal high precision independent clock source, internal & external watchdogs.

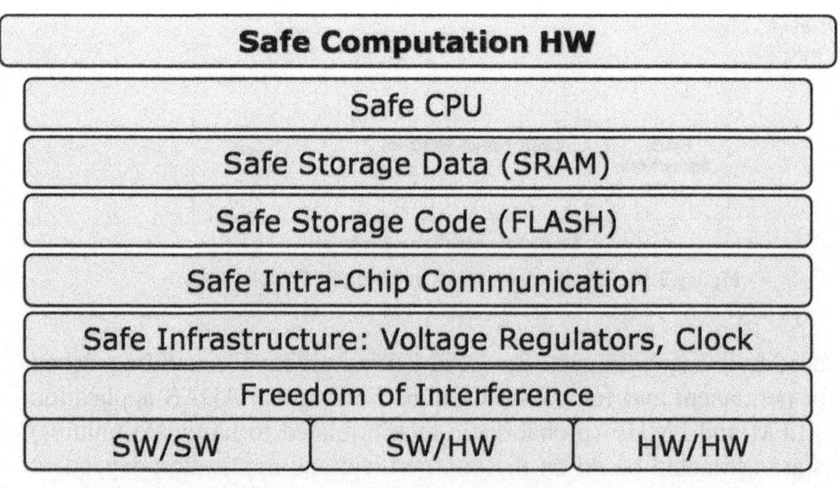

Figure 2.19 Generic elements of safe computation hardware platform.

Finally power supply range monitoring is implemented for the internal regulators.

To achieve a safe computation software platform an ASIL D compliant operating system needs to be used featuring memory protection and time protection. Further it needs to provide services for program flow monitoring, end-to-end communication safety protocols as well as safe interrupt vector generation. ASIL D compliant software is required to be developed according to ISO 26262 part 6.

The AURIX platform ensures freedom of interference at software level by means of SW isolation, while freedom of interference at hardware level is guaranteed by HW isolation. The CPU MPU (memory protection unit) monitors the direct access to the local memories, applies to software tasks and allows dynamic re-configuration. The bus MPU monitors the SRAM accesses via interconnect. Finally register access protection monitors write access rights to module registers.

References

[1] DESERVE deliverable D1.2.1 – Development platform requirements.
[2] DESERVE deliverable D1.3.1 – Development platform specification.
[3] DESERVE deliverable D1.3.2 – Method and tools specifications.
[4] DESERVE deliverable D2.5.2 – Platform system architecture.
[5] DESERVE deliverable D2.5.4 – Standard interfaces definition.
[6] AUTOSAR, http://www.autosar.org
[7] ISO 26262, Road vehicles – Functional safety (www.iso.org).
[8] A. Sandberg, D. J. Chen, H. Lönn, R. Johansson, L. Feng, M. Törngren, S. Torchiaro, R. Tavakoli-Kolagari, A. Abele – Model-based Safety Engineering of Interdependent Functions in Automotive Vehicles Using EAST-ADL2, Lecture Notes in Computer Science, Volume 6351, Series: Computer Safety, Reliability, and Security (SAFECOMP), Pages 332–346. Springer Berlin/Heidelberg, 2011. ISSN 0302-9743.
[9] www.interactive-ip.eu
[10] www.haveit-eu.org
[11] S. Durekovic (NAVTEQ), Perception Horizon: Approach to Accident Avoidance by Active Intervention, Workshop "How can new sensor technologies impact next generation safety systems?" IEEE IV 2011, June 5 2011, Baden–Baden.
[12] DESERVE Deliverable D2.2.1 – Perception layer Preliminary Release.

[13] AURIX Safety Manual, Infineon confidential document, no. AP32224, v1.1, dated Sept. 2014.
[14] K. Östberg und M. Bengtsson, "Run time safety analysis for automotive systems in an open and adaptive environment," in SAFECOMP 2013 – Workshop ASCoMS (Architecting Safety in Collaborative Mobile Systems), Toulouse, France, 2013.
[15] J. Frtunikj, M. Asmbruster und A. Knoll, "Data-Centric Middleware support for ASIL assessment and decomposition in open automotive systems".

3

Driver Modelling

Jens Klimke and Lutz Eckstein

Institute for Automotive Engineering, RWTH Aachen University, Steinbachstraße 7, 52074 Aachen, Germany

3.1 Introduction

Traffic simulations become more and more relevant for the development of Advanced Driver Assistant Systems (ADAS) and algorithms for automated driving. They are used to evaluate the functions concerning important impact factors like safety, efficiency, mobility or costs. Therefore, the system is tested and evaluated as a component of the virtual vehicle in simulations. The factors manageability and acceptance of the users regarding the tested system are prospected and evaluated in driving simulators, where the real driver can be part of the virtual environment. Both, in traffic simulations and in simulators, the realistic behaviour of the surrounding virtual road users to the equipped vehicle is an important requirement for a suitable evaluation of the system because this behaviour influences the reaction of ADAS and driver significantly. Moreover, it is necessary, that the behaviour of the traffic can be adjusted systematically in order to generate defined traffic situations of relevant constellations and in different nuances of criticality. As in real traffic, small changes in the initial conditions can produce a large difference in the result. This phenomenon can only be reproduced in a simulation if the driving behaviour patterns reflect the human driver behaviour closely.

The basis of this *driver model* and its possible functionality or ability is the underlying simulation environment. To determine the risk of congestion for example, a traffic simulation environment with *macroscopic*, e.g., fluid dynamic based traffic behaviour, is suitable. The easiest macroscopic representation of virtual traffic could be an equation with the result of an average velocity dependent on the density of traffic. This might be a complex mathematical relation producing suitable results for some purposes but it is impossible to understand the specific inner traffic effects like congestion waves and traffic collapses. For such effects, the influences of the traffic elements on the driver models have to be understood.

These are basically the interactions between the driver-vehicle-units among each other and the reactions of the units to the traffic environment like traffic light systems or the road curvature. In this kind of traffic simulation, called *microscopic* traffic simulation, the desired controlling reaction of the driver or the automated function is calculated and implemented directly into the vehicle. This is done in form of a change of the dynamic state of the vehicle, e.g., a desired acceleration, which consequently results a change of velocity and position. The driver and the vehicle represent an inseparable unit, but entirely with a unit-specific behaviour. The behaviour might respect some dynamic restrictions of the vehicle and in some cases of the driver, but does not depict the driver-vehicle-interaction.

For the analysis of modern ADAS this kind of simulation is not suitable, as a driver has, e.g., to be able to override the system by using the control elements, like pedals, steering wheel or switches. An ACC for example can be switched off in critical situation by using the brake pedal or can be overridden by using the accelerator pedal to further increase or keep the acceleration. These effects can only be simulated if vehicle and driver are implemented as separate models and if the interfaces between driver model and vehicle model are used to implement the driver's wish to the vehicle. Thus, this concept can be called *sub-microscopic* or *nanoscopic*.

Another specific application for sub-microscopic traffic simulations is the exploration of detailed effects related to the vehicle, like fuel consumption analysis in specific traffic situations or environments. Within these analyses a very detailed vehicle model is needed. But it is not only the specific application which let us chose a higher level of traffic simulation. Obviously, the higher the level of detail, the more effects can be depicted with a single traffic simulation environment and model set-up but at the expense of computing time up to the loss of the real-time capability. Additionally, the effort of setting up the models increases due to the increase of model parameters. For the same reason the validation of the models is much more complex, too.

In the past decades many driver models where developed with special focuses on different specific elements of the driving task. Some try to show an optimal behaviour, without taking into account the physical and cognitive abilities and limitations of the human driver. Others focus on these restrictions or on the information process in the driver's brain and body and the capability of the driver to process different information in parallel. In literature many categories of driver models are published. Jürgensohn defines in [1] two basic categories of driver models, formal and non-formal models. *Formal* models have a fixed description but a changeable inner value. The result of formal models is reproducible, that means, the same conditions lead to the same output. *Non-formal* models are not described by those fixed dependencies (like equations or lingual definition) or they have a non-changeable (constant) character. Examples of formal models are *descriptive* models, which have a fixed description but have a character which is not defined by an input-output structure. In the European research project ASPECSS [2] and in Deliverable D3.1.1 [3] of the DESERVE project the definition is different. In these sources *descriptive* models are clearly defined (fixed, but not constant) and generate a numeric, quantitative output dependent on different numerical influences. This output is reproducible but can anyway contain stochastic elements. *Functional* models describe physical and psychological aspects of driving, like the information processes, the human structure of thinking and acting. They do not generate a numeric output but draw a picture of the elements of driving. The difference between functional and descriptive models in this definition is not unique and not complete; there are hybrid models and models which can't be matched to any of these categories. In this chapter, the distinction between *formal* and *functional* models is used to avoid the conflict of the two definitions of *descriptive* models.

In complex traffic simulations the usage of both kinds of models is needed to depict realistic traffic flow and driving behaviour. Formal models describe algorithms for a driver model how to reach its goal by setting defined reference values dependent on the input. Functional models can help to understand the driver's wishes and to create an eligible structure and decision algorithm.

In the DESERVE project, a rapid prototyping platform for the development of ADAS was created and a suitable tool-chain for the development process was outlined. The traffic simulation is an important tool in the development process of ADAS and thus is part of the DESERVE tool-chain. As described above, a realistic driver model is needed for the development and evaluation of modern ADAS. In the next sections, the way of modelling the driving behaviour is described, followed by the requirements for the DESERVE driver model. On the basis of the requirements the structure of a sophisticated driver model is developed and the used implementation techniques and strategies are explained. In the last section two different applications of the driver models are presented.

3.2 Driver Modelling

Driving is not just a single decision and a single action at once. It is rather a complex interoperation of different motivations, perceptions, decisions and states with continuous and discrete changes. To create a realistic driver model, a strict delimitation between these elements has to be done and it is helpful to create a suitable structure with a unique and logical naming of the elements and well-defined interfaces. To develop such a structure, driving has to be analysed on the basis of typical driving scenarios, manoeuvres and actions.

Besides the perception and the handling or action, the information processing is the most important part of driving. Within the information processing, the driver estimates desired values for different future vehicle states he wants to achieve, like a desired speed, a desired following distance, and distance to stop. These inner desired states are called driver-variables or briefly *variables*. Often a driver has multiple desired values for the same variable, generated by different motivations, between which a decision is needed. As an example the desired speed shall be used: The driver can have multiple causes of choosing a desired speed. For example the following three: First, to reach the destination as soon as possible. Second, the speed limits on the road. Third, the curvature of the road combined with the need for safety. For each motivation, a desired speed can be determined. The speed limit for the first mentioned motivation is the maximum speed the driver would choose on a free, straight road. If there are no further influences like other road-users or speed limits, the driver would travel with this speed. Situations, which do not allow travelling with this speed, do not imply that it is not the driver's wish (the driver wants to, but can't). For the second motivation, a speed in an interval around the speed limit, dependent on the law-abiding is desired. This can be higher or lower or exactly the speed limit. The third motivation results in a desired speed which allows the driver to pass a curve in a comfortable and safe manner.

All described motivations lead to different speeds, so the driver is in a dilemma: She/he has to decide for one speed to accelerate or decelerate to. The decision in this case is taken in a pragmatic way: The lowest speed wins, because on the one hand there is a comfort and safety limit, on the other hand there is a limit because the driver accepts the given speed limits or at least wants to avoid fees for driving too fast.

The described example shows two input types to the driving behaviour, the driver's character (here: need for safety, need for comfort and law-abiding) and the current situation described by the state of the own vehicle and other vehicles as well as the road and environmental structure. Moreover, not only the local situation influences driving. A good driver reacts before approaching to a discrete situation to reach the desired value in time. In the curve speed example above, a real driver would estimate the comfortable and safe speed based on the visual perception of the road's curvature before reaching the curve. On that perception, the driver decelerates with a rate which leads to the desired speed at the moment the curve is reached. Within the curve the driver corrects this estimation to satisfy the desired safety and comfort. The predictive behaviour is called *anticipatory* driving. The correction is called *compensatory* driving [4]. This phenomenon also has to be regarded in the development of driver models.

Of course the driver has more responsibilities than the decision of the desired speed. According to Rasmussen [5], the driving task can be seen in three levels: The strategic level where the driver plans and creates strategic values like a route, the manoeuvring level, where the driver processes the decisions and determines desired values and value sequences, the strategy can be implemented with. This behaviour is conscious: The driver knows exactly how to solve the driving task and creates a strategy. The driver is able to reflect decisions and actions he/she took in this level. In the control level the driver implements these conscious values into the vehicle by using the steering wheel, the accelerator and brake pedal and other control elements of the vehicle. This operation is not done in a single step. Often the driver determines a subconsciously desired value, like a desired acceleration, which is then transferred into the actual vehicle input. This value is not reflected by an experienced driver. It is an automatism by the driver to reach the conscious desired value. The desired speed shall be used for an illustration: After the decision to move freely, because no other road-user is influencing the driver, the desired speed is detected, which is a conscious value. To reach this speed, the driver accelerates with the desired acceleration, which is a subconscious value because the driver cannot quantify this value and it is not part of the strategy. The final implementation is done by using the vehicle's controls to reach this acceleration. The advantage of using this subconscious step is that the regarded values can be set, manipulated and limited dependent on realistic driver's needs independent of the conscious behaviour. Often the desired acceleration and yaw rate or curvature is used as an output of macroscopic driver models. In this definition these variables represent subconscious variables. Thus, without the implementation by using steering wheel and pedals, the model can be seen as a macroscopic driver model.

3.3 Requirements for DESERVE

Before creating a driver model, an analysis of the requirements for this model based on the field of application has to be done. In DESERVE, a rapid prototyping platform and development process has been created. The details of the platform can be found in Chapter 2. This requirements section concentrates on the applications of the DESERVE platform. In the first year of the project, the needs for the driver model were analysed in D3.1.1 [3]. There are two kinds of driver models identified in the project: the virtual driver for the usage in traffic simulations like described above and the driver intention and distraction model, which is used as a component of an ADAS to detect the real driver's state.

The literature review, the analysis of existing driver model concepts and in particular the research work in the DESERVE project shows that it is not possible to create one holistic driver model to satisfy all scientific needs. Nevertheless it would be very attractive, if there was one basic structure combining the ideas of the previous research, in which the algorithms can be added as independent modules. The connections of all modules – with properly defined affiliation and interfaces and in conjunction with a suitable parameter set – will produce the expected results. For that reason, a generic module-based structure needs to be developed which is well-defined and flexible for amendments. Most of the integrated algorithms can be used for several applications while others are specific to one. The generic structure should fit to all applications of driver modelling in an open way.

Another important issue is the implementation. Many driver models are implemented in native programming languages. This fact has a significant disadvantage: It becomes very muddled due to the one dimensional structure of programming code. Often driver model structures are shown in a two dimensional representation with levels in the up-down dimension and sequence of the information processing in the left-right direction (time related). An implementation of the driver model in an analogous structure could be very helpful to create a clear and well-arranged model. Thus, a graphical implementation would be aspired. Furthermore, it should be possible to structure or capsulate the content properly as well as the definition of the interfaces to take the advantage of modern programming techniques like object oriented programming or code reuse to avoid redundancy. Next to the structural requirements, the system shall be able to hold values or states over one or more time steps to implement the memory of the driver. Another requirement is the possibility to connect the driver model to the traffic simulation environment. This can be done by communication interfaces or by the native integration of the compiled driver model, for example as a dynamic linked library or similar techniques.

The driver model (virtual driver) in DESERVE shall be used in different traffic simulation environments for testing and evaluating ADAS functions in the process of the development. Within the project, the driver model shall be implemented and tested for a control function which is designed to show the advantages and benefits of the DESERVE platform. Therefore, an Advanced Cruise Control system (ACC) is combined with a Heading Control (HC). The system shall assist the driver on inter-urban road scenarios and increase the safety within the full speed range (WP 4.2, [6, 7]). The decision for demonstrating the system for the inter-urban area is made, because this area is a very important research field for the usage of ADAS functions of the next generation; especially those who reach the next level of driving automation (cf. SAE automation level 2 – partial automation, [8]). Also the evaluation of ADAS for the increase of safety is important in the inter-urban area. Therefore, detailed driver models are needed with the claim to be valid for the intended purpose. In particular, the modelling of realistic human behaviour on intersections and junctions is one of the most important developments for today's traffic simulations in order to develop ADAS with the goal to reduce the high number of accidents on intersections.

Analysing the application in DESERVE, the driver model requirements can be briefly defined:

- Inter-urban driving behaviour including safe-passing of slow, right-moving vehicles has to be implemented.
- The driver model needs the capability of route-following within multi-lane roads and complex but flexible transport networks.
- Full intersection and traffic light behaviour has to be implemented.
- Anticipatory driving behaviour, like early speed adaption needs to be reflected.
- Re-use of validated driving behaviour algorithms and driver model approaches is required.

The driver model is implemented and connected to the simulation environment PELOPS [9]. The inter-urban ACC and HC developed in DESERVE is tested in virtual traffic scenarios containing units controlled by the here described driver model. These scenarios include straight and curvy multi-lane roads, complex intersections with traffic lights and right-of-the-way controls by signs and structure, different speed limits, rare and dense traffic with different parameterisations and slow moving vehicles (e.g. mopeds). This testing set-up leads to a set of manoeuvres and primary driving tasks which have to be implemented:

Longitudinal	Lateral
Free moving	Lane keeping
Approaching/Following	Curve cutting
Braking in critical situations	

Figure 3.1 Primary driving tasks which are implemented in the driver model within the DESERVE project separated by longitudinal and lateral control.

Longitudinal	Lateral
Stopping, Standing and starting	
Turning on intersections	
Lane change	
Safe passing	

Figure 3.2 Manoeuvres which are implemented in the driver model within the DESERVE project.

There are several other manoeuvres which can be implemented like U-turning or stopping on the road side. These manoeuvres are not implemented within DESERVE. Nevertheless, the structure of the model shall offer the possibility to enhance the functionality.

3.4 Generic Structure

In this chapter the ika driver model is introduced. Within the DESERVE project, a suitable and generic driver model structure was developed and implemented which fulfils the requirements from the previous section. The interfaces and driver parameters are defined and described in this chapter.

3.4.1 Model Structure

From literature review, two generic structures can be identified: The three levels of driving by Rasmussen and the three blocks of perception, information processing and action, which can be found in several formal and non-formal model approaches (e.g. [10]). This leads to a matrix-form model shown in Figure 3.3. The modules (blue boxes) in the matrix represent model implementations or parts of those. The arrows, in different shades of grey, describe the information flow between the blocks and represent the internal

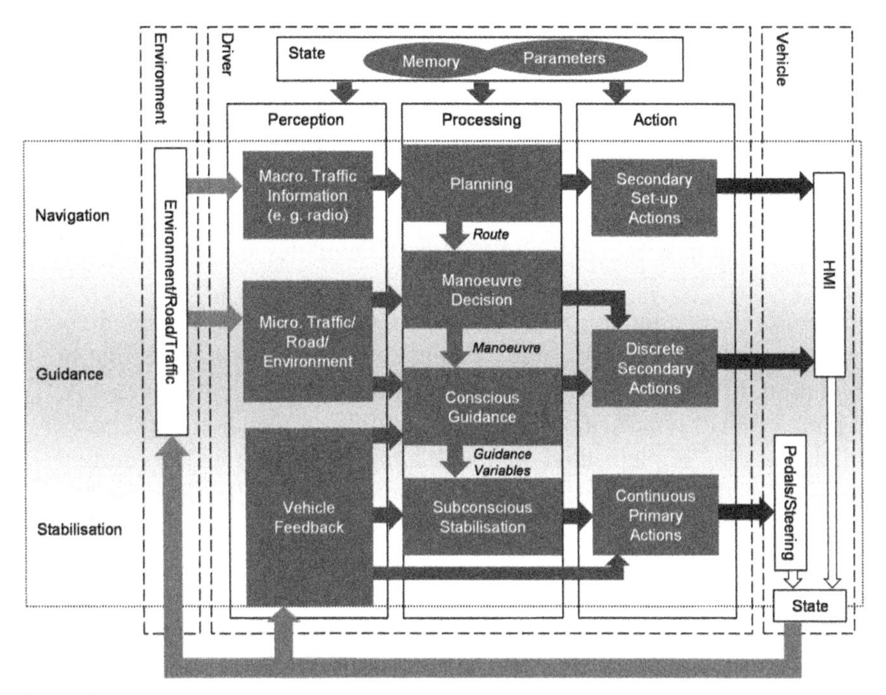

Figure 3.3 Driver model structure in the context of environment and vehicle: the structure includes perception, processing and action blocks including its functional modules and the regarded dynamic information flow.

interfaces. The blue arrows show the information flow through the three levels and represent the needed information (variables) for the driving tasks and manoeuvres. A central functional block of the model is the *State* block, where the driver-specific values are stored. The *Memory* module represents the driver's knowledge about the current situation, the manoeuvre states, the destination or route, etc. The memory is used to keep information for the following time steps, during the manoeuvre or for the whole simulation cycle. This information can be extrapolated to estimate current states of the ego-vehicle or other road-user even if the driver model does not sense the regarded information at the current time step. Thus, the memory has an interface to the *Perception* block and constitutes an input of this block besides the inputs of the environment and the vehicle. Current manoeuvre states and important values, which have to be known in the next time step, are also saved in the memory and are passed by the interface between the *State* and the *Processing* block where the driving calculation is implemented. The parameters represent the driver's character and are defined in two layers: qualitative and physical parameters (see Subsection 3.4.2). The *Parameters* block serves its values to all blocks of the driver model, for example by manipulating the handling time delay (reaction time). The *Action* block controls the handling or the conversion of the driver's wish into physical actions like the manipulation of the pedals, the steering wheel, shifting and using the HMI control elements.

As it can be seen in Figure 3.3, a strict assignment of all modules to a unique level is not possible. In the following the modules shall be explained in detail.

In the *Planning* module, route specific calculations are executed. In general, the units have a fixed route calculated or set in the initialisation of the simulation. In reality a driver changes the route under circumstances, e.g., traffic jams or road blocks. If such functionalities are needed, appropriate algorithms can be implemented in the Planning module. In the current implementation the route is stored in the memory. The Planning module calculates a value for each lane in the environment around the unit, which gives a quantitative value of how far the lane and its successors can be followed on the given route. Thus, the Manoeuvre Decision module can decide which lane the driver wants to take. The *Manoeuvre Decision* module processes all discrete manoeuvres and discrete decisions. That means on the one hand to decide for a manoeuvre and on the other hand to control the manoeuvre but not to calculate the related Guidance Values. The Decision module returns different states within the manoeuvre and process variables, which can be used by the following modules to perform the manoeuvre (in the figure briefly named *Manoeuvre*). An example is described in Section 3.5. Another output of the module is a set of *Discrete Secondary Actions* which are needed or desired at the beginning or during the manoeuvre. This can be for example switching the turning indicators in case of turning or lane changes. On the basis of the decision with its states and process values, a local strategy to perform these manoeuvres and continuous driving tasks is calculated in the *Conscious Guidance* module. Continuous driving tasks are performed during the whole simulation time without the need of a discrete decision. Of course, the output values of these tasks can be overridden by other results. An example is the motivation to keep the lane: This task is continuous because the driver always wants to stay in the lane but can be forced to leave the lane during an overtaking manoeuvre. Within the Conscious Guidance module the *Guidance Variables* are filled with values (guidance values), which the driver wants to reach. An example was given in Section 3.2 (desired speed during free moving). Several guidance values are calculated and passed to the *Subconscious Stabilisation* module. Within this module, desired stabilisation values are calculated.

In general, these values are the desired acceleration and the desired yaw rate for the longitudinal and lateral control respectively. Based on all motivations the stabilisation value with the highest benefit for the driver is taken. Besides the desired values, some real physical values, which are states of the vehicle, can be directly sensed by the driver. Thus, the driver is able to implement these values subconsciously by using the vehicle control elements (pedals and steering wheel). This implementation is done in the *Continuous Primary Actions* module.

To define the interfaces between the modules it is helpful to create a manoeuvre and driving task table. For the DESERVE implementation the following tables (Figure 3.4 and Figure 3.5) were developed, derived from Figure 3.1 and Figure 3.2.

In the motivation of *free moving*, the desired velocity of the driver is calculated. This velocity depends on the speed limit, the curvature of the road ahead and the maximum desired velocity of the driver. To reach the velocity, the driver model accelerates (subconsciously) dependent on the current velocity and the desired velocity. A suitable model approach is part of the Intelligent Driver Model (IDM) by Treiber, Hennecke and Helbing in [11]. An adaption of that approach for the usage in complex driving simulations is published in [12]. The *following* motivation is mainly influenced by a desired following distance which bases on a driver specific following time gap. To reach this distance the driver needs to accelerate or decelerate. The *lane-keeping* is performed by the usage of fix-points based on the Two-Point Visual Control Model published in [13]. This model can be adapted, so that the fix-points cause a yaw rate, which the driver wants to implement. The adaption is published in [14]. The yaw rate is chosen as the desired subconscious stabilisation value because it physically implies both, the curvature and the velocity. During standing, the driver model maintains a brake pedal value which results in a vehicle that does not move. This means that the pedal value is a subconscious value, different to the other longitudinal tasks.

Motivation	Conscious	Subconscious	Action
Free moving	Velocity	Acceleration	Pedal value
Following	Distance	Acceleration	Pedal value
Lane keeping	Fix-points	Yaw rate	Steering wh. angle
Standing	-	Pedal value	-

Figure 3.4 Process variables for the four basic driving motivations *free moving*, *following*, *lane keeping* and *standing*.

Manoeuvre	Conscious	Subconscious	Action
Lane change (Two-phase lc.)	1. Lat. offset 2. Fix points	1. Lat. velocity 2. Yaw rate	Steering wheel angle
Stopping	Stop point	Acceleration	Pedal value
Safe passing	Fix-points	Yaw rate	Steering wheel angle

Figure 3.5 Process variables for the three manoeuvres *lane change*, *stopping* and *Safe Passing*.

In Figure 3.5, the manoeuvre *turning* is missing. In this model turning is implemented in the decision module, at least to control the turning indicators, but does not require a process implementation due to the given features: A lateral and longitudinal turning manoeuvre can be seen as a 'normal' street following motivation if the turning path is known and a turning speed is calculated by the given curvature. In the case of conflicts with 'right of way' road-users (e.g. at left turns), the driver model stops with the manoeuvre *stopping*. If the conflict is resolved, the stop manoeuvre is aborted, so the driver model switches to *free moving* or *following*.

The perception is partly done in the simulation environment: All perceived information is transformed to the driver's coordinate system by the simulation environment. The driver model adapts the information with driver specific perception errors, like perception limits, continuous noise, sporadic disturbances or fluctuations and accuracy limits.

3.4.2 Parameter Structure

In many driver model approaches, *physical parameters* are used to influence the driver behaviour and generate heterogeneous or driver specific results like in the IDM [11]. Examples of physical parameters are the maximum comfortable acceleration and deceleration or a constant following time gap to the leading vehicle. These parameters are well measurable for a single driver or a group of drivers, represent a direct input to the model approaches and are mostly independent of each other. To describe the character of a driver, a big set of physical parameters has to be defined. In other driver models *humanised parameters* on a higher level are used which are not directly measurable. These parameters have a meaning which can be described as a characteristic or a constant attribute of a human driver. In general, the parameters are used to generate driver specific physical parameters, which are then dependent on each other by this humanised characteristic. With these parameters a characterisation of the driver is easier because the number of parameters is reduced to a smaller number. The challenge is to create a mathematical dependency which returns realistic results based on these fictive parameters. The humanised parameters used in the driver model for the DESERVE platform are named *sportiness*, *need for safety*, *law-abiding* and *estimation ability*. However, these parameters have no scientific physical or psychological meaning; they only represent groups of drivers and influence the underlying parameter block of physical parameters like *desired following time gap*, *acceleration profile* and many more. In Figure 3.6, the parameter concept of the DESERVE platform is shown: In the first block, the humanised parameters are shown. These parameters influence the physical parameters of the driver model. In this example, the *need for safety* parameter influences the *lower and upper following time gap* (see [15]) and the *acceleration profile* of the driver model. Parameters are not influenced by the dynamic inputs.

The set-up of a suitable parameter concept influencing all models in a realistic way is difficult and extremely dependent on the implemented model

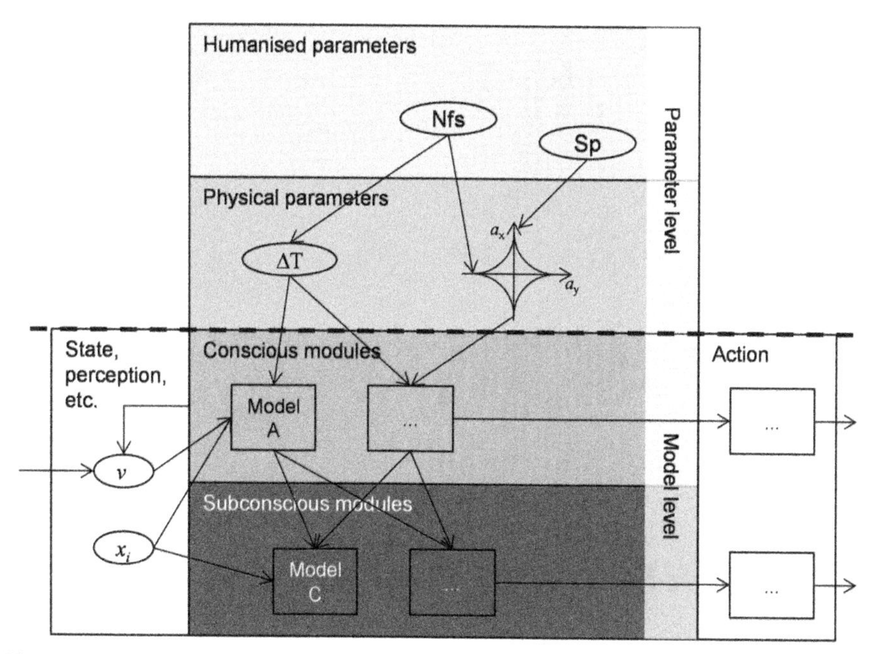

Figure 3.6 Sketch of the parameter blocks (brown) and model blocks (blue) of the driver model.

approaches. A concept to solve this problem could be to measure a large set of reference data and run an optimization to find the best fitting parameters. After that a validation has to be done with another set of data to prove the concept.

To create a traceable connection between the parameter blocks, in the DESERVE model, cubic polynomial functions are used. In a review of floating car data for example, the distribution of lower following time gaps of the Wiedemann model was generated. Basis of the distribution of these time gaps is a Gaussian distribution of the need for safety parameter with $\mu = 0.5$ and $\sigma = 0.15$ as described in [15]. With the polynomial

$$\Delta T_{\text{lower}}\left(p_{\text{NFS}}\right) = 1.4 \cdot p_{\text{NFS}}^{3} + 0.9 \cdot p_{\text{NFS}}^{2} + 0.9 \cdot p_{\text{NFS}} \tag{3.1}$$

with

ΔT_{lower}: Lower following time gap [s]
p_{NFS}: Need for safety; Gaussian distributed (0.5, 0.15) [–],

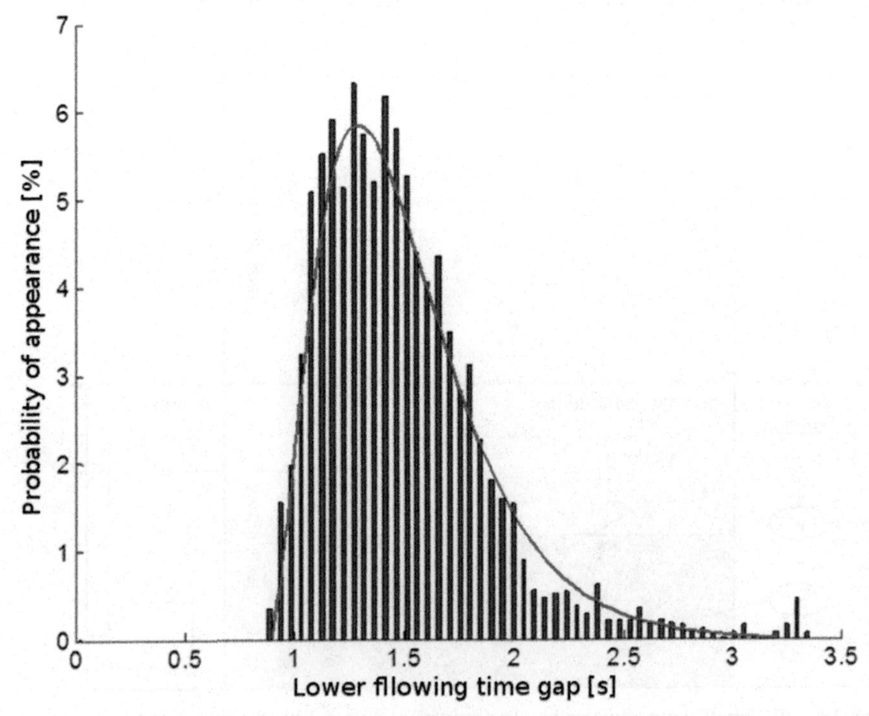

Figure 3.7 Distribution of lower following time gaps for real drivers (blue bars) and the modelled distribution dependent on a normal distributed *need for safety* parameter (red line).

the distribution of the lower following time gap returns a result shown as red curve in Figure 3.7. The blue bars show the floating car data which is the basis of the polynomial curve in this example.

This principle can be used and optimised analogously for the other physical parameters.

3.5 Implementation

The graphical programming tool *Matlab/Simulink* provides the implementation features described in the requirements in Section 3.3. The 2D graphical GUI allows a clear and well-arranged implementation close to the visual structure of the model. The implementation is easy to understand and easy to debug. In the university environment, many students and scientific assistance work with the driver model for a limited time range (e.g. Bachelor/master theses or PhD theses). Thus, a further important requirement is the comprehensibility of the model. Programming in Simulink is easy to learn also without deep knowledge of classic programming languages. The code can be capsulated in subsystems with defined inputs and outputs and several storage concepts can be used to implement the driver's memory. The data connection between the model and other tools can be established by using UDP or TCP/IP or other versatile techniques.

For the DESERVE example implementation, PELOPS is used as the simulation kernel with the support of environmental structures (road network, traffic lights, etc.) and vehicle models. The core of the new version of PELOPS is implemented in Java. The integration of a Simulink model is possible with the UPD communication interface. For the simulation of one vehicle this solution is suitable and is real-time capable in the current version of the ika driver model and PELOPS. If multiple vehicles use the same driver model instance with their specific inputs, at least time-dependent and memory-containing modules do not work properly. For the simulation of at least two vehicles, the Simulink-model needs to be duplicated to have an independent copy (second instance) of the driver model. This becomes difficult for a high or flexible number of vehicles in a simulation. Another problem is the high execution time due to the UDP connection and the Simulink model itself. A native execution combined with direct data exchange, e.g. by shared memory, is much faster. The Matlab/Simulink tool-chain brings the possibility of code generation: The desired model can be converted to C or C++ code which can be integrated in other C/C++ or FORTRAN code or can be compiled to a shared library in almost all computing platforms. In DESERVE this solution is used to integrate the driver model into PELOPS. For that purpose, a class wrapper is used around the generated code. That allows the simulation environment to create almost infinite numbers of independent driver model instances. Multiple test cases have been performed to show the capability of running traffic simulations with the full functionality of the driver models and a large number of traffic units in real time.

Except for the decision module, all modules are implemented in standard Simulink subsystems with mathematical blocks. The decision module is implemented in *Stateflow*, which is an integrated Simulink feature. Stateflow allows implementing state machines, which is a suitable implementation technique for discrete decision structures. To demonstrate a possible implementation of a manoeuvre decision the lane change shall be used as an example: In Figure 3.8, a state machine implementation is shown for a lane change decision including the *progress* and *sequence* control. The progress describes the state or the 'position' in the lane change like *initialisation (init)*, *origin lane*, *lane crossing (LC)*, *target lane* and *termination (term)*. The phases describe the phase control of the lane change by the driver. In this example the driver uses two phases to perform the lane change: In the first phase the driver accelerates laterally to a desired lateral velocity (anticipatory) dependent on the lateral offset. In the second phase, the driver 'switches' to the lane-keeping mode with the focus on the target lane (compensatory) by using the fix-point approach (see Figure 3.5). Dependent on the phase and the progress, the conscious guidance module, calculates the reference values which are needed to steer the vehicle to the desired lane. The transition A denotes the decision to perform the lane change, which is valid if there is a lane next to the ego driving path with

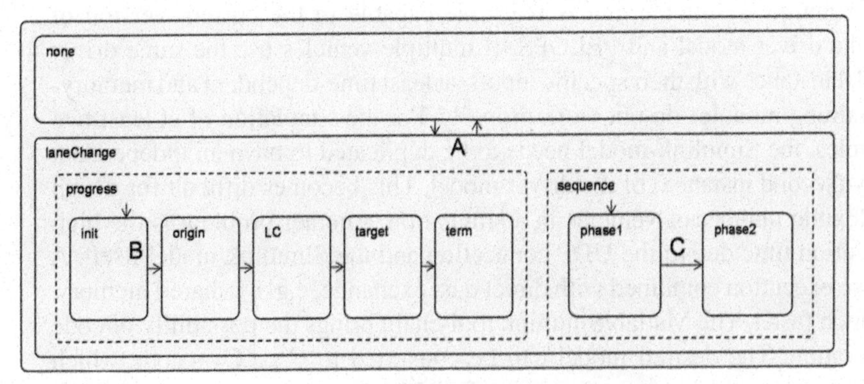

Figure 3.8 Stateflow model for a two-phase lane change including decision (A), progress control (B) and sequence control (C).

higher correlation to the route and some other conditions, like distance to the end of the lane, preference lane and a hysteresis. The basis for the decision is described in [16]. A decision for a lane change does not mean an immediate reaction. The driver model can decide before the lane or the desired gap is reached. In the case of a positive decision, the lane change is initialized. This is a continuous process as long as the active lane change is not started. The transitions B control the progress of the lane change and transition C represents the transition from the first phase to the second one in this example.

3.6 Applications in DESERVE and Results

Within the DESERVE project, the driver model was used for two different applications: The validation of left turn simulations within the full parameter range and the prediction of a real driver regarding the acceleration during free driving, approaching and following.

For the validation of left turn simulations (in this example without stopping), real traffic data from laser scanners were used to measure the trajectories of 136 vehicles on a junction in Alsdorf, close to Aachen in Germany. Figure 3.9 shows the results of the simulations for different parameter sets (coloured curves). The measured real-driver data are shown

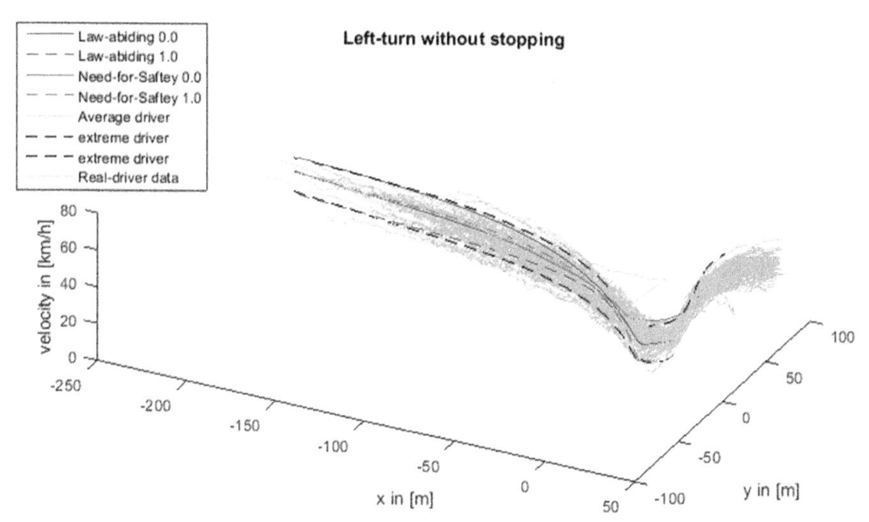

Figure 3.9 Trajectories (velocity over x- and y-position) for left turn including the simulation results for different parameter sets. The real driver data is measured on one intersection with 136 different drivers during day time.

in grey and the boundaries (extreme driver) as well as the average driver are included. The extreme drivers are generated by choosing respectively, the maximum and the minimum, of the need-for-safety and law-abiding parameters. For this example, the upper extreme driver is created by setting the need-for-safety and the law-abiding parameters to zero and the lower extreme driver is created by setting these parameters to one. As it can also be seen in the figure, the law-abiding parameter influences the speed the driver reaches before and after passing the intersection but not the velocity during the turning (red lines). Opposite to this, the need-for-safety parameter influences the speed within the turning only (blue line). This result depicts the statement that the turning speed is mainly driven by the safety and comfort motivations of the driver and the speed on straight roads is defined by the acceptance of speed limits. The phases between approaching and turning are representing a mixture of all motivations and result in a transition of the speed. In this example, the other parameters are set to the average value (0.5).

To predict the driving behaviour of a real driver in a vehicle, the driver model was integrated as a module on a real-time system in the car, equipped with real sensor data by radar and camera sensors. A five second simulation is calculated in each prediction step and the result is written to the CAN-Bus. With that data ADAS like ACC can react dependent on the estimated wish of the driver. The system and the results are published in [12].

3.7 Conclusions and Outlook

In the DESERVE project a driver model structure was developed with the focus on the realistic generation of driver-vehicle-environment interactions. For the usage in traffic simulations the driver model has been implemented in Matlab/Simulink and exemplarily been integrated in PELOPS. The addressed traffic area covered the inter-urban road network including generic intersections. Therefore, common driver model approaches but also conceived approaches to create the modules needed in DESERVE were used to obtain realistic driving behaviour. The elementary interactions between the driver models, the associated vehicles and the surrounded environment result in realistic traffic phenomena and effects occurring in equivalent real traffic situations which was shown by comparing the simulation results with measured data on a real intersection. The model behaviour is tuneable via parameters on two levels, a humanized and a physical level, which have indirect and direct influence on the model behaviour.

The structure of the model was designed to offer the possibility of enhancing the driver model by using different model approaches or expanding it with the capability of performing yet unimplemented manoeuvres and driving tasks. In those cases, the challenge is to tune the added model approaches while maintaining the realistic influence of the parameters. To simplify and partly automate the tuning process a tool can be implemented which uses real data to optimize the mathematical influence of the parameters to the model. This work will be done in the future to increase the usability of the driver model for the simulative analysis of traffic situations. The traffic simulation and thus the driver model shall be an inherent part of the tool chain used in the development of ADAS and functions of automated driving.

References

[1] T. Jürgensohn and K.-P. Timpe, *Kraftfahrzeugführung*. Berlin, Heidelberg, Springer, 2001.

[2] D. Raudszus, M. Ranovona, S. Geronimi, M. Kunert, E. Schubert, and T. Schaller, "Report on Driver and Pedestrian Reaction Models", Project Deliverable, ASPECSS, 2013.

[3] S. Fruttaldo, G. Piccinini, D. Pinotti, R. Tadei, G. Perboli, L. Gobbato, A. Zlocki, J. Klimke, F. Christen, N. Pallaro, F. Palma, and F. Tango, "D3.1.1 – Standard Driver Model definition", Project Deliverable, DESERVE, 2013.

[4] E. Donges, "A two-level model of driver steering behavior," *Human Factors,* Vol. 20, No. 6, Dec 1978, pp. 691–707, 1978.

[5] J. Rasmussen, "Skills, rules, and knowledge; signals, signs, and symbols, and other distinctions in human performance models," *IEEE Transactions on Systems, Man and Cybernetics*, Vol. SMC-13, no. 3, pp. 257–266, 1983.

[6] J. Klimke, F. Christen, N. Pallaro, A. Kyytinen, P. van Koningsbruggen, E. Nordin, and X. Savatier, "D4.2.1 – Control functions solution design", Project Deliverable, DESERVE, 2013.

[7] J. Klimke, F. Christen, and L. Eckstein, "Definition of a Microscopic Traffic Simulations Driver Model for Inter-urban Intersections for 21st World Congress," in *ITS World Congress 2014*, Detroit, 2014.

[8] SAE International, *Taxonomy and definitions for terms related to on-road motor vehicle automated driving systems*. SAE International Standard J3016, 2014.

[9] "PELOPS Whitepaper," Forschungsgesellschaft Kraftfahrwesen Aachen mbH (fka), Aachen, 2014 http://www.fka.de/pdf/pelops_whitepaper.pdf.

[10] L. Eckstein, *Active Vehicle Safety and Driver Assistance Systems, Automotive Engineering III*. Lecture Notes, Institute for Automotive Engineering (ika), Aachen, 2015.

[11] M. Treiber, A. Hennecke, and D. Helbing, "Congested Traffic States in Empirical Observations and Microscopic Simulations," Rev. E 62, Issue, Vol. 62, p. 2000, 2000.

[12] J. Klimke, P. Themann, C. Klas, and L. Eckstein, "Definition of an embedded driver model for driving behavior prediction within the DESERVE platform," in *International Conference on Embedded Computer Systems: Architectures, Modeling, and Simulation (SAMOS XIV), 2014*, 2014, pp. 343–350.

[13] D. D. Salvucci and R. Gray, "A two-point visual control model of steering," *Perception*, Vol. 33, No. 10 (2004), p. 1233–1248, 2004.

[14] J. Klimke, C. Klas, and L. Eckstein, "Konzept zur Strukturierung eines generischen Fahrermodells anhand des realen Informationsflusses," in *VDI-Fortschritt-Berichte: Reihe 22, Mensch-Maschine-Systeme*, 2015.

[15] R. Wiedemann, *Simulation des Straßenverkehrsflusses*. Karlsruhe: Institut für Verkehrswesen, 1974.

[16] D. Ehmanns, *Modellierung des taktischen Fahrerverhaltens bei Spurwechselvorgängen*. Dissertation, Institute for Automotive Engineering (ika), Aachen, 2003.

4

Component Based Middleware for Rapid Development of Multi-Modal Applications

Gwenaël Dunand

Intempora, France

4.1 Introduction

Developing multi-modal applications starting from scratch is a tough issue. On the one hand, there are algorithms challenges such as detecting drowsiness or pedestrians in every possible situation. On the other hand, there are programming challenges such as handling multiple sensors data with different frequencies and different nature (video streams, GPS data, laser scans, etc.), as well as implementation details, such as synchronization techniques, multithreading and memory management, for only naming a few.

Moreover, the time required to develop the software is often underestimated [1]. Using an already existing middleware helps to keep on schedule and focus mainly on business problems while decreasing the real-time programming complexity.

There are several middleware that fit all those previous descriptions (ADTF, PolySync, BaseLabs and RTMaps). As RTMaps is the official middleware chosen for the DESERVE project and the author is very familiar with this one, this chapter will sometimes be focused on RTMaps, but other tools might apply as well.

4.2 Using a Middleware

Considering software as layered, middleware incorporates many of these layers vertically. A middleware provides a full, or partial, solution to an area within the application and supplies more than the basic library, it also supplies associated tools like logging, debugging and performance measurement.

Because middleware is vertical system, it may compete or duplicate other parts of the application.

4.3 The Multisensor Problem

The number of sensors used for ADAS applications has increased in the last few years. Now applications use radars, lidars, GPS, high definition stereo cameras, lasers, IMU, CAN Bus, eye trackers, V2V and V2I communication, etc... The problem is how to read all of them within the same application and especially how to synchronize them despite their very different nature (Figure 4.1).

As a matter of fact, most algorithms need to use several sensors to reach a good level of detection. The problem is that those sensors might have different sampling rates, or even worse, event-based outputs. Reading from those sensors simultaneously can be a tricky problem to solve. Let's illustrate this with an example with three signals.

In the Figure 4.2, signal **A** (orange) and signal **B** (green) are periodic with a different period while signal **C** (red) is an event-based signal. One solution would be to use the least common denominator of all sampling rates to perform the reading. While this approach may work with periodic signals like A and B, it won't work with the event-based C signal.

To achieve reading from multi-modal sensors, RTMaps middleware is fully **asynchronous** – each component runs in its own thread – so that any

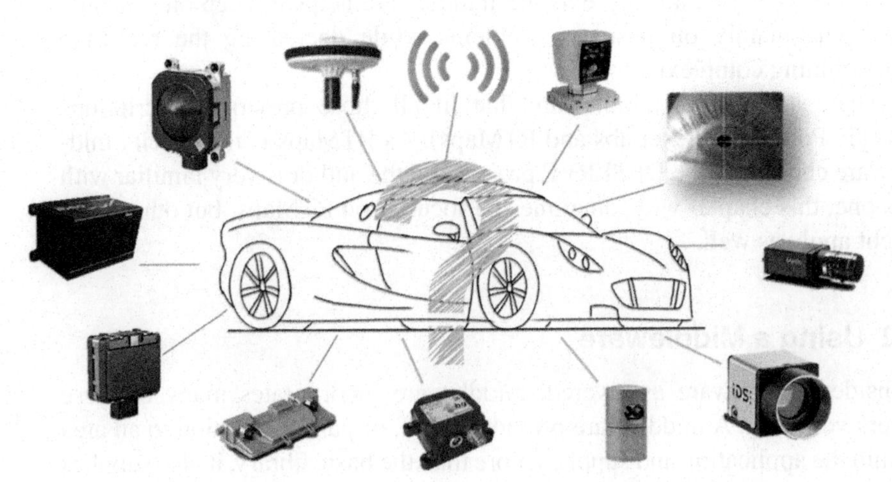

Figure 4.1 ADAS function requires many different type of sensor.

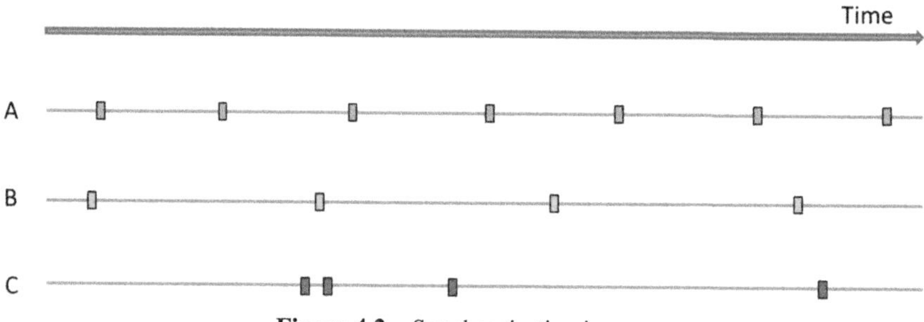

Figure 4.2 Synchronisation issues.

component can react to any data stream, whatever sampling rate it may have. This is the only way to follow the natural pace of each data. This design uses internally blocking calls, removing any extra latency that could happen when using polling methods. RTMaps middleware also defines reading policies to synchronize data streams. While the default policy – *reactive* – works perfectly fine in most case, the user can use one of those:

- *Reactive reading*: a component with multiple inputs will read every time a new data sample is made available on any one of its inputs.
- *Synchronized reading*: a component with multiple inputs will process one sample from each input when data sample with the same timestamps (plus or minus some configurable tolerance) are available on its inputs. This behaviour is made for data fusion and allows re-synchronization of the data streams at any point downstream in the diagram, whatever the latency of the various upstream data channels.
- *Triggered reading*: a component with multiple inputs will read when a new data sample is made available on a given input. It will then resample the data on its other inputs through non-blocking reading.

To sum-up, not only the middleware provides a common platform to build the ADAS application, but it also does take care of the tricky data synchronisation mechanism.

4.3.1 Knowing the Date and Time of Your Data

Using a middleware allows to be very accurate about the timing of your data. For example, RTMaps affects two timestamps to the data: the *timestamp* and the *time of issue*.

- The *timestamp* is the intrinsic date of the sample. It is as close as possible to the date of occurrence of the real data which the sample corresponds to. It is often supplied by the first component that created the sample (i.e. the acquisition component). The *timestamp* remains unmodified while the sample goes through the different components of the processing chain. The *timestamp* often corresponds to the date where the data is available in system memory.

- The *time of issue* is the date corresponding to the last time the sample was output from a component. Therefore, this date increases as long as the sample runs through the different processing components.

Knowing with precision the time and date of your data is essential to perform synchronized readings (*see previous section*), but it is also useful to estimate the latency of your data or know the processing time of a component which is really vital in real-time applications.

4.3.2 Component-based GUI

RTMaps middleware comes with a user-friendly graphical interface which allows building an application using components (seen as blocks) connected to each other. The Figure 4.3 shows RTMaps studio with a diagram open and a few components in it.

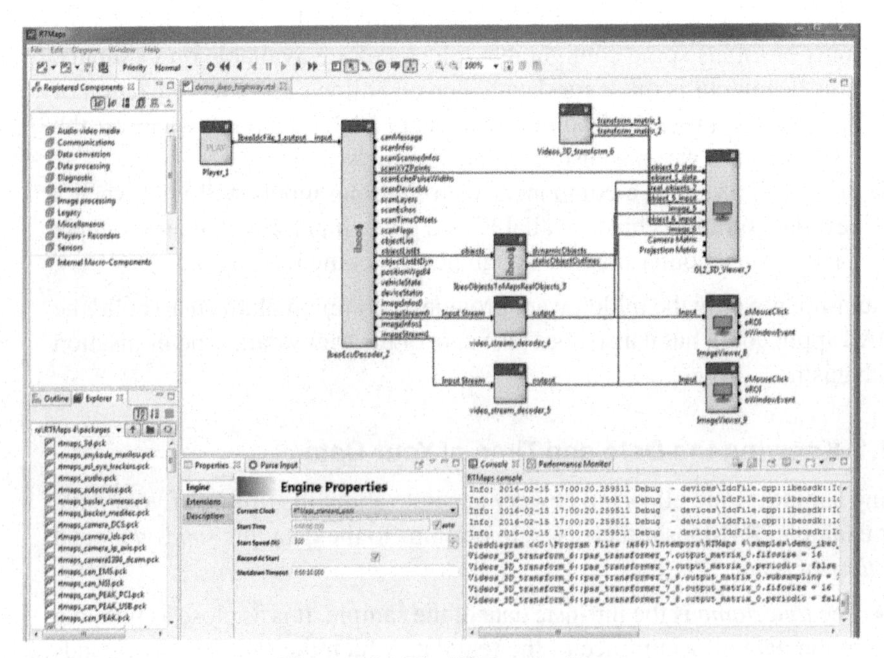

Figure 4.3 The RTMaps Studio.

The advantage of using a graphical user interface is twofold. Firstly, it allows the user to quickly construct an application by using drag and drop techniques and wiring components to each other. Realizing a simple demonstration with a camera and an IMU only takes a few minutes [2] whereas using only hand-written code with dedicated libraries would take weeks.

Secondly, it allows the team to focus on interfaces. This is a very important point since it defines boundaries and clarifies the work between teams. In big projects like the DESERVE project, strict definitions about interfaces are necessary due to the number of partners. The interface for components is composed of inputs, outputs and properties. Once the interface of a task is defined, changing an algorithm for another is not a problem anymore, one component can be replaced by another and the work is done! In the Figure 4.4, the face detection component has one fixed detection interface. The input is YUV image and the output is a vector of rectangle representing the faces found.

Furthermore, the use of macro-components can definitely simplify the diagram by splitting the global problem into sub-problems (Figure 4.4). All the implementation is hidden in first appearance to simplify the reading, but of course looking under the mask would reveal all the internal details.

4.3.3 The Off-the-Shelf Component Library

The off-the-self component library represents all the already available components in the middleware. This is an important part of it because it allows accelerating the application development by using and reusing already developed component. Here are a few categories of components:

- *Sensor interface*: This category represents all the components that allow to read/write from/to a sensor. When a sensor is present in the library, the user has just to drop a corresponding component on the current diagram and configure it to retrieve the data. That work can be done easily with a consequent time benefit.

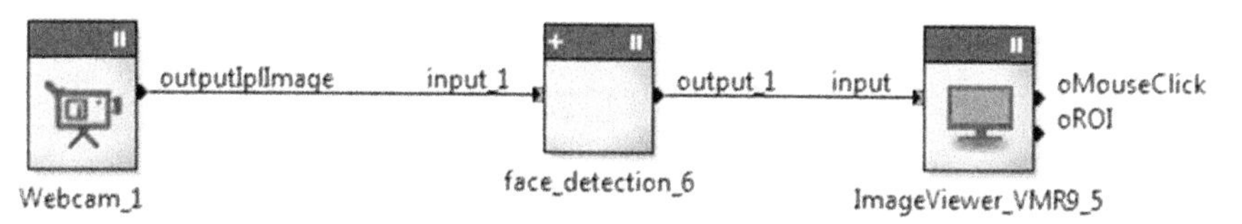

Figure 4.4 Components and interfaces.

- *Data generators*: When comes the time of testing a component, it might be useful to emulate a missing sensor with random generated data (vectors, CAN frames, images). This does not replace real sensors but it can be enough sometimes.
- *Viewers*: Very important libraries, which allow displaying information about data stream during the execution (images, vectors, CAN frames...). As an example, the *DataViewer* (Figure 4.5) can display generic information (timestamps, size, etc.) and specific ones (width and height of an image if current data is an image) as a tree. This is very useful to inspect data along a processing chain and check that such component behaves correctly.
- *Player and Recorder*: Those components allow to record and replay any data stream. Using a recorder, the user is able to record any scenario (outdoor session, motorway driving test, automatic car parking, etc.) and replay it at the office with the exact same data and timestamps.

DataViewer_13

Data	Value	
Randint_14.outputInteger		
Info		
Data		
[0]	10	
RandomStream8_15.oStream8		
Info		
Time of issue	25001519	
Timestamp	25001519	
Vector size	3	
Buffer size	10	
Type	Stream8	
Data		
[0]	145	(0x91)
[1]	115	(0x73)
[2]	8	(0x08)
[3]	74	(0x4a)
[4]	189	(0xbd)
[5]	185	(0xb9)
[6]	255	(0xff)
[7]	83	(0x53)
[8]	75	(0x4b)
[9]	211	(0xd3)

Figure 4.5 Inspecting data with the data viewer.

4.3.4 Custom Extensions

Extending the component library is done through the **SDK**, whose purpose is to expand the capabilities of the middleware by the creation of new components. In RTMaps for example, the SDK is available for both C++ and Python (Figure 4.6). Thanks to this SDK, the user can integrate his own code into a component and use it directly in this diagram.

Once a new component has been created, it can be shared with others. When using C++, each component is compiled code which means that only the binary code is used in the middleware and so the IP is preserved. Anybody can share his work while keeping the source secret.

4.3.5 About Performance

Using a high performance middleware is still essential nowadays. Indeed, even if the power of the computer tends to increase continuously, the trend is to run applications on embedded systems with the smallest footprint possible. The explanation of this trend is quite simple: the prototype vehicle has to be as close as possible as the real vehicle. In many companies, no desktop computer

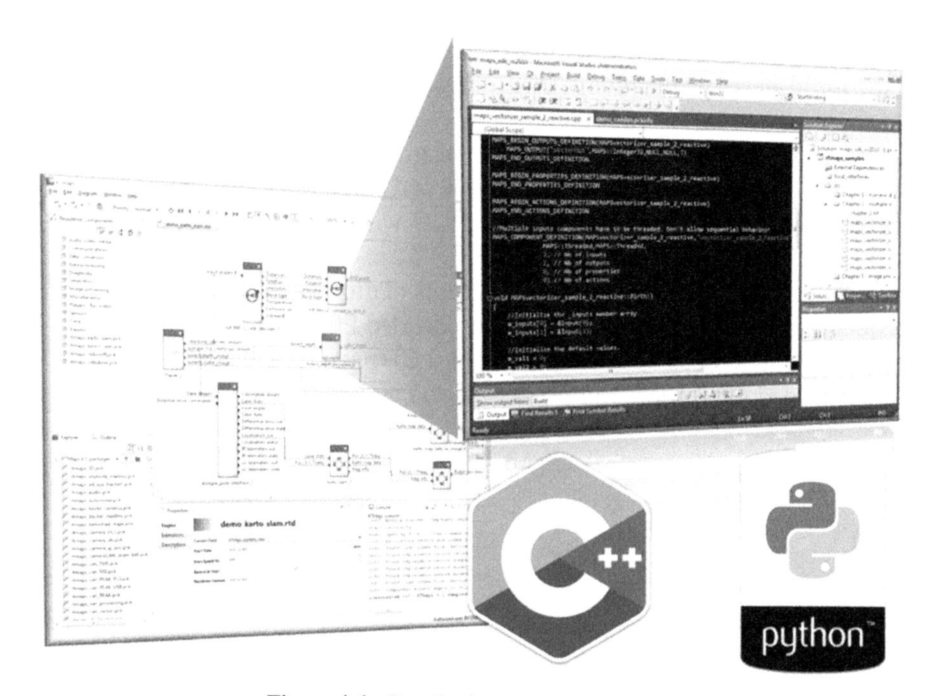

Figure 4.6 Developing a new component.

in the trunk of the car are allowed anymore, all systems have to be (or at least look) embedded.

Furthermore, the middleware is pushed further and further in the development chain. A few years ago, most of the middleware were assigned to do only prototyping and once the prototype application was finished, all the work had to be done again on dedicated hardware. This not the case anymore, now the middleware should be able to run on low consumption cards that equip pre-series cars.

Consequently, OEMs are looking for high performance middleware that runs on small form factor cards as well as on Personal Computer so that working on lab or real scenarios makes no difference.

4.4 Compatibility with Other Tools

4.4.1 dSPACE Prototyping Systems

In the frame of the DESERVE project, a bridge has been developed between the dSPACE MicroAutoBox and RTMaps (Figure 4.7). The dSPACE MicroAutobox is the de facto standard for real-time control loop such as chassis control, body control and powertrain. Combining this dSPACE prototyping system to the RTMaps middleware provides an extremely powerful framework capable of doing multisensor acquisition, data processing and controlling actuators in a hard real-time way.

The MicroAutoBox typically serves as an embedded controller to process the ADAS application algorithms in real-time and to interface the vehicle bus, sensors and actuators. It is a prototyping ECU with a predefined set of I/O which is qualified for in-vehicle use.

In the context of the DESERVE project this platform was extended by an Embedded PC and an FPGA Board. The embedded PC features a multi-core Intel® Core™ i7 processor running at 2.5/3.2 GHz and the connection to the actual embedded controller is implemented via an internal Gigabit Ethernet

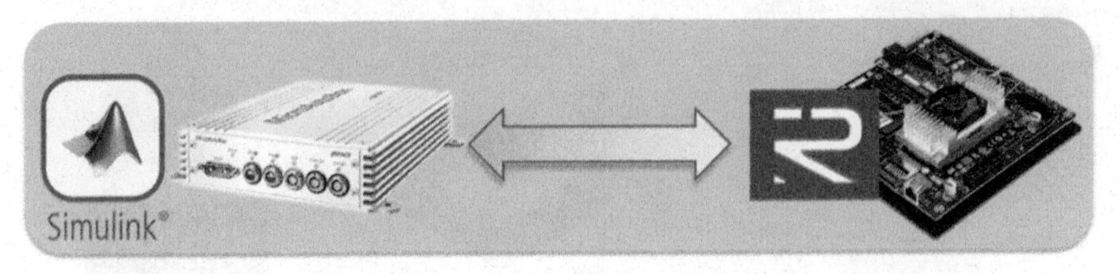

Figure 4.7 dSPACE MicroAutobox and RTMaps Bridge.

interface. The embedded PC integrated in the MicroAutoBox can be used to flexibly run any x86 based development framework available for prototyping perception and fusion algorithms, such as RTMaps, and to exchange easily data with the embedded controller [3].

4.4.2 Simulators

ADAS are becoming more and more promoted because several key functions permit to increase the level of vehicle safety. Most of the time, it is a challenge to access to the equipment and sensors information on vehicles, making difficult to design and test these new algorithms. Some of the applications are based on perception sensors embarked on the vehicle, which interact with the vehicle, driver and environment through electronic control units. For those reasons, the simulations of the algorithms and the analysis of existing solutions for virtual testing are very important tasks.

Using simulators has many advantages: tune the scenario at will (add rain or fog like in Figure 4.8), test dangerous situations where real data is hard to get, use the output of any algorithm to modify the scenario of the simulator (close the loop), etc. It's pretty much a fact now; virtual testing allows massive

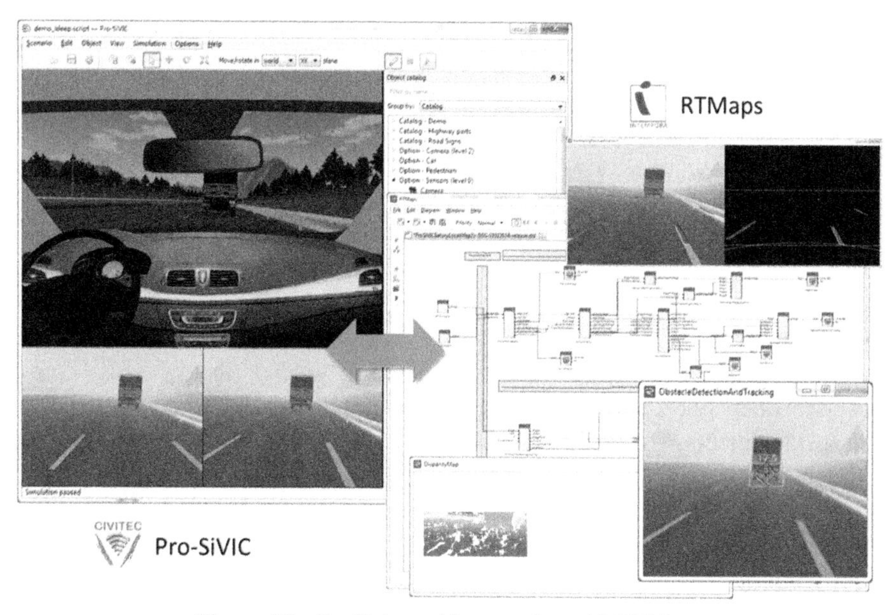

Figure 4.8 ProSivic working together with RTMaps.

reduction cost. In the DESERVE project, many simulators have been used in collaboration with RTMaps: ProSivic [4], dSPACE ASM [5], etc.

4.4.3 Other Standards

Middleware supports other standards as well. RTMaps implements the DDS [6] standard interface via the Prismtech OpenSpliceDDS implementation. This is very convenient to stream data from RTMaps to anywhere and vice-versa. This DDS interface was developed in the frame of the DESERVE project.

Other standard protocols are also supported, like XIL or XCP, which allow manipulating RTMaps with off-the-self tools that implements those protocols themselves.

Of course, most of the middleware on the market will also support NMEA, CAN/DBC, RTSP, I2C, GPS, SIP, TCP and UDP as well. The compatibility with major industry standards is essential so that the middleware interacts painlessly with other tools.

4.5 Conclusion

Most DESERVE partners have been using RTMaps and ADTF middleware as the common perception platform to speed up their development processes and exchange components between each other.

Indeed, partners like Continental, FICOSA, Vislab and CTAG have encapsulated their acquisition routines and custom algorithms into RTMaps components, which in turn have been integrated into a global acquisition and processing diagram by other partners (OEMs most of the time). This modular approach made the collaboration easier between a large number of partners, which was one of the difficulties of the DESERVE project.

Another example, CRF (*Centro Ricerche Fiat*) has used RTMaps and the bridge to the MicroAutoBox – developed in the frame of the DESERVE project – for their emergency breaking application. The sensor acquisition, the pedestrian detection, information display and the breaking order are done *via* RTMaps.

As a conclusion, in the DESERVE project, having a middleware has allowed engineers to focus on their main activity – obviously ADAS functions here – and not on advanced programming issues, but it was also very helpful to exchange components between partners.

References

[1] Software Engineering 8th Edition, p. 109, ISBN-13: 978-0321313799, 2006.
[2] Intempora. (2012, February 20). *RTMaps4 demo* [Video File]. Retrieved from https://www.youtube.com/watch?v=HBxFq04S91g
[3] Joshué Pérez Rastelli, David Gonzalez Bautista, Fawzi Nashashibi, Fabio Tango, Nereo Pallaro, et al. Development and Design of a Platform for Arbitration and Sharing Control Applications – a DESERVE approach-. IEEE SAMOS Conference, Jul 2014, Samos, Greece, pp. 322–328.
[4] Prosivic. (2016, June 21). Retrieved from http://www.civitec.com/
[5] dSPACE. (2016, July 12). *Simulation tool suite*. Retrieved from https://www.dspace.com/en/inc/home/products/sw/automotive_simulation_models.cfm
[6] OMG. (2016, July 11). *DDS: the proven data connectivity standard for the IoT*. Retrieved from http://portals.omg.org/dds/

5

Tuning of ADAS Functions Using Design Space Exploration

Abhishek Ravi[1], Hans Michael Koegeler[1] and Andrea Saroldi[2]

[1]AVL List Gmbh, Austria
[2]C.R.F. S.C.p.A , Italy

5.1 Introduction

An ADAS function developed within the DESERVE platform and the tuning of this function for a particular application is discussed in this chapter. Based on separating the software and tuning data, according to the standards described in detail in Chapter 2, such a function can also be used for an alternate vehicle or application use case. The opportunities as well as the potential challenges are described, using a real world example, developed within the DESERVE Project.

5.1.1 Parameter Tuning: An Overview

Tuning or calibration of vehicle components is essentially determining the optimum attributes, which fulfill the legislative standards as well as refine the car's character to meet all the expectations of the driver for drivability and comfort. Besides the comfort and legislative issues the vehicle tuning also helps in brand differentiation and helps to determine the vehicle character.

In the tuning task for a specific component (e.g.: engine), the software and the tuning data in the application layer of an Electronic Control Unit (ECU) is separated which is illustrated in Figure 5.1. The resulting code is a hex file, which can be flashed to the defined controller hardware which gives a big flexibility in powertrain development. As an example, one engine hardware can be put into more than 200 vehicle variants fitting for different countries, different vehicles and/or different transmission systems – just by flashing a different appropriate controller software.

Figure 5.1 Separation of software and tuning parameters in a control unit.

5.1.2 Industrial Tuning Applications: Challenges and Opportunities

The engine – ECU has been the first mechatronic application in the automotive world. It makes sense to have a short view on the historical development of the tuning task in this field as illustrated in Figure 5.2.

In the past decades, the improving technology in the automotive sector can be seen with cars having better engine performance, less consumption, better handling and reduced emissions. But the improvement in technology has come with increased complexity, especially in the tuning task.

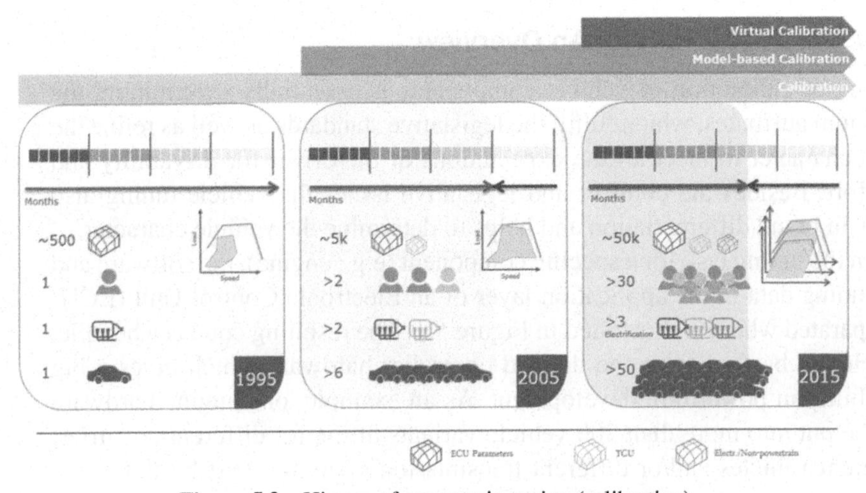

Figure 5.2 History of powertrain tuning (calibration).

As can be seen in Figure 5.2, initially there used to be around 500 parameters which needed to be tuned, which was carried out by a single engineer using the unit to be tested, which was then tested on a single test vehicle. Initially, the powertrain was quite simple and the Engine – ECU was the only one being considered.

With increasing legislative and user demands; the complexity of the technology, the number of involved interacting components (engine, gearbox and electric engine) and also the number of functions controlling the interactions between all the variable components increased dramatically. Further the tuning allowed the derivation of many more vehicle variants with the same hardware components but differing in the ECU-SW, wherein the functions in the SW stay the same, just the tuning data are specifically developed.

This effect is also seen in the number of tuning parameters to be defined in an engine calibration project, where around 50 k parameters have to be defined – clearly assigned to many functions. So it is no longer possible to have *one person*, who understands all the functions implemented and teams of specialized persons are necessary, partly working in different areas of the world. Thus the industry was confronted with several challenges and found some responses.

For example, the management of tuning data becomes an issue. It must be possible to track all the changes made to the tuning data by the different engineers involved and bring all the tuning results into a single final tuning result. The company should be able to ensure at Start of Production (SoP) that:

1. All the tuning data are calibrated.
2. All the tuning data are calibrated with the correct settings to optimally fulfill the desired, derivative use case.

These two requirements are very challenging, which explains the need of "Tuning Data Management". This topic itself is not further elaborated in this chapter, but is supported by valuable literature [1, 2].

Another challenge lies in the tuning for single use cases: For example, the emission tuning of an engine in a certain vehicle configuration for the legislation of a specific country. There are about 5 to 10 strongly interacting tuning parameters. E.g. an engine map to define the start of the combustion as function of speed and load is counted as one of these parameters, and exhaust gas recirculation rate, rail pressure, boost pressure, split patterns of the injected fuel quantity are others, all either reducing the different kinds of emissions or changing fuel consumption or noise.

So one can imagine, that it is just not possible to measure the emissions and the fuel consumption of all the feasible combinations of say 8 of such parameters on an engine. (A similar issue faced with ADAS functionality)

Such tasks are typically performed on engine test beds and chassis dynos and have to be finally validated on the road again. With the latest legislation (Real Driving Emissions, RDE) even the certification will be done on the road giving additional challenge [3–5].

Figure 5.3 illustrates the generalized development environment, which allows the engineer to reproduce maneuvers and then double check the results of tuning work. In the manual tuning method, the engineer operates the UUT with a certain setting of control parameters in certain maneuvers. The engineer observes the behavior of the UUT and performs a judgment according to his experience. Then the next setting is defined with the intention to better approach the desired behavior. This process becomes complex when there are many relevant tuning parameters [6].

In this trial and error method, the quality of tuning and the optimization results depend on whether the engineer considers all the parameters that are relevant for the desired behavior and the relevant start point. There is a strong dependence on the experience of the engineer. There are also limitation on the number of tests that can be conducted, due to the testing time, complexity and cost factors. The final results are highly subjective, as the decision making

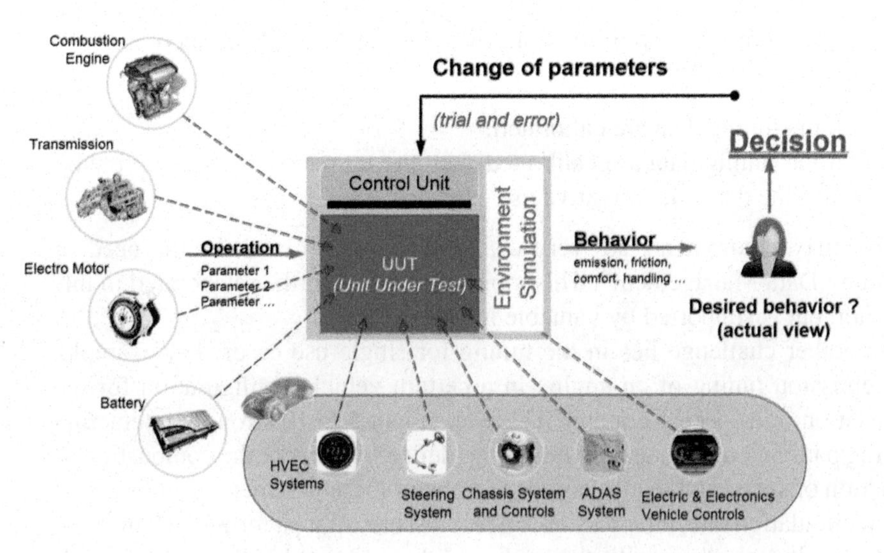

Figure 5.3 Illustration of a generalized development environment and manual tuning process.

process lacks traceability and a reuse is not possible for future projects, e.g. tuning an ADAS setup for a different drive mode. As a result, a methodology to increase the efficiency and the quality of the tuning work at the same time, the so called "Design of Experiment" method (DoE) was adapted accordingly.

Within the DESERVE context this methodology was applied as "Design Space Exploration" for Simulation environments, which are excellent development environments for tuning of ADAS Functions.

The model-based approach was used with two objectives:

- Firstly, to find an optimum tuning result.
- Secondly, to validate an existing tuning result under a big variety of use cases, which will happen during the lifetime of a vehicle.

5.1.3 Model-based Tuning

Model-based tuning is a statistical, model-based approach which reduces the amount of actual experiments/test runs needed to accurately describe the behavior of the UUT within the design space. This method helps to choose the position of the test data points in order to generate behavior models with an efficient low number of measurements. Such models are then utilized to develop an accurate and robust tuning according to specific optimization target(s). In Figure 5.4 the entire method is illustrated again for the generalized development environment.

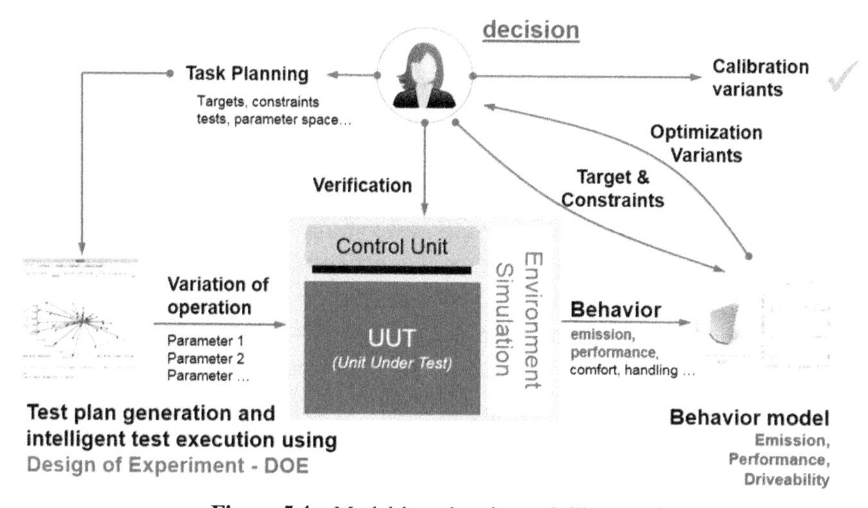

Figure 5.4 Model-based tuning task illustrated.

In a model-based tuning task the below steps are followed:

- The user begins with a task planning for the measurement series, where the targets for the tuning task are determined. Based on the targets, the relevant input parameters which are considered to influence the observed UUT response are selected. AVL CAMEO is used for the test plan generation. This is based on a one time set up process, in which CAMEO is connected to the development environment. Thus CAMEO gets access to set tuning parameters in the UUT, observe responses of the UUT and to start/stop maneuvers and to take measurements after maneuver. The development environment hosting the UUT could be in the form of a test bed, a hardware-in-the-loop (HiL) or even a vehicle simulation software like IPG Carmaker in combination with an ADAS-function prototype programmed in MATLAB.
- Once the targets have been defined the next important step is to make the test matrix. In order to get a full picture of the area to be investigated, the Design of Experiments (DoE) is used [7]. It is a systematic technique which allows varying all the parameters simultaneously while answering the two important questions of every tuning activity: Firstly, how many tests are needed to cover the entire design space? And secondly, at which locations in the design space test points are needed to effectively get modelling equations valid throughout the entire design space. There are many DoE designs available to us in AVL CAMEO, but COR DoE methodology [8] was used in the current example exercise. Besides setting up the test design, it is also important to set the limits for the test and appropriate actions when the limit is violated. These topics are addressed further on in the example discussed in Subsection 6.2.1.
- With the test plan and limits decided the tests are run, where the necessary parameter settings are uploaded to the UUT by CAMEO, and after the test, the required measurement results were stored in CAMEO. The raw measured data check is then carried out in order to check the plausibility and feasibility of measurement. It is a necessary check to get a rough idea of how the measurements compare against expected values, and also observe possible errors which could have occurred during the test execution.
- The measurements are modeled empirically to obtain behavior models of the UUT. In this content, modeling means more or less to fit a function – like a polynomial equation for example – into the measured responses in order to estimate the response function of any point in the design space.

Such a model helps understand the reaction of the UUT to the parameter tuning, and the interaction of the different tuning input parameters and the output measurements. The confidence and prediction intervals of the empirical models are observed to evaluate the model quality. Models in CAMEO also allow extrapolation in defined ranges beyond the design space covered by measurements to observe the UUT behavior at points where tests could not be run based on equipment limitations or time/cost constraints.

- Based on the optimization target, optimization algorithms can be implemented for a single objective or multiple objectives. The engineer can decide if the results meet the targets and constraints and in case of multiple objectives decide on a suitable tradeoff between the different desired targets (Pareto front).
- Before, the results from the analysis are accepted a final verification test is carried out. Tests are run at least on the point of the decided optimum, but can also be extended on parameters settings of ten or more points spread across the Pareto front. If these verification measurements match the modeled results then the empirical models are accepted and the engineer can use the optimization results as the desired tuning setting.

5.1.4 Model-based Validation

A model-based validation is a task carried out to test and evaluate the robustness of the results from the tuning task. The UUT is run at the parameters settings obtained from the tuning task, but tested for an alternate use case and the response is evaluated. For example; if say a diesel engine was tuned to operate at an economy mode and a sport mode with strong limits set on NOx emissions. Economy mode encourages the engine to conserve fuel while sacrificing power, while the Sport mode encourages the engine to provide greater power while making compromises on fuel economy, with the engine running more at the higher RPMs. The engine is initially tuned at driving conditions imitating an urban environment and lower altitudes, and from the tuning tasks the input parameters settings like the rail pressure, injection pressure, injection timing etc. are selected to operate the engine at the two targeted modes while sticking to the NOx limits. In the validation test run the engine is first run at the economic mode and then sport mode, but now the use case is in hilly road conditions and higher altitude. The engine performance is evaluated with respect to power and emissions, while the road and altitude of operation is varied. The target is to see if tuning settings could be extrapolated or extended to alternate use cases. It also gives further information on how the engine tuned for urban conditions would perform on rugged hilly conditions.

5.2 Demonstrative Example

A map-based ACC-Function (developed by the DESERVE Partner CRF) running in a commercially available MiL Environment (IPG-Carmaker + MATLAB Simulink) has been used as an example. The calibration tool of AVL CAMEO was connected to this environment in order to tune the function for a Fiat 500L.

5.2.1 Function: An Overview

A map-adaptive autonomous cruise control (ACC) was developed to:

- Control the vehicle velocity in order to enter and exit curves in a comfortable and safe manner.
- Complete the drive maneuver in the least amount of time.

The controller function controls the vehicle speed by sending jerk request (see Figure 5.7). Jerk is the rate of change of acceleration. Hence the jerk request signals from the controller function are converted into the vehicle acceleration and speed. For the reference maneuver a digitized road was used and a reference speed curve was determined, which is the maximum speed at which this road can be safely maneuvered. The function tries to ensure that, the vehicle follows this reference speed profile as closely as possible without exceeding it. The target speed was set at 130 km/h for the ACC. A demonstrative speed profile is shown in Figure 5.5 for a sample settings in

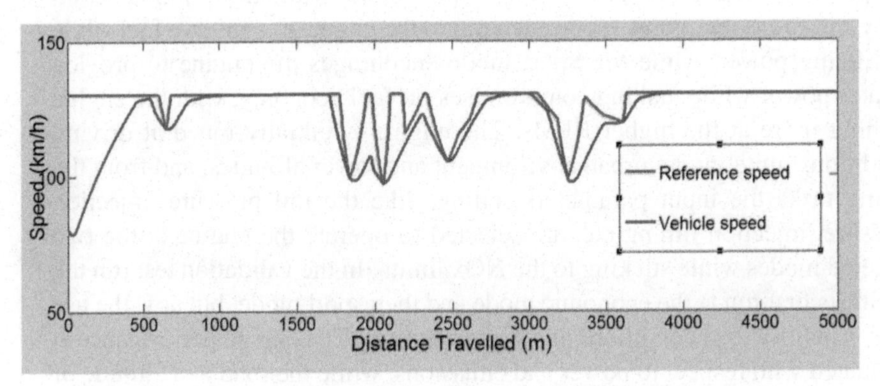

Figure 5.5 Velocity profiles for a sample test run using the control function.

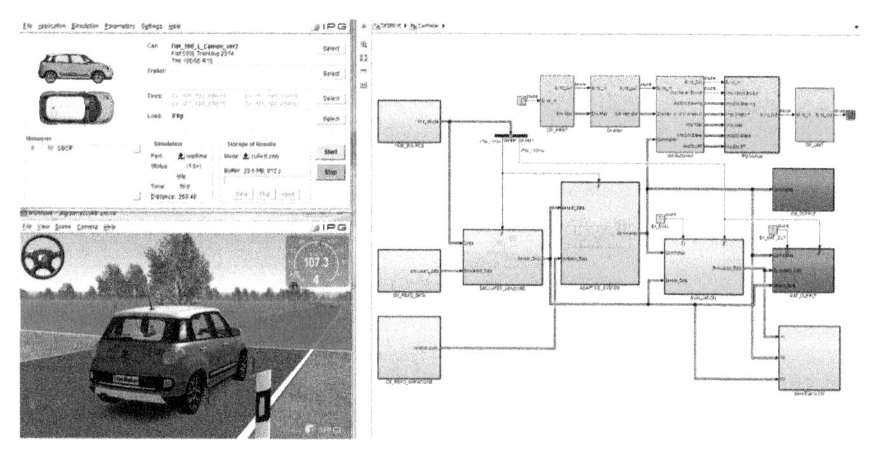

Figure 5.6 Function developed using IPG carmaker and MATLAB simulink.

the ACC function. It can be seen that the vehicle velocity tries to follow the reference velocity while never exceeding it. The vehicle velocity is not able to exactly replicate the reference velocity due the road conditions, the vehicle limitations and the control function settings.

The function was developed using IPG Carmaker for Simulink and has been illustrated in Figure 5.6. IPG Carmaker for Simulink is integrated into MATLAB/Simulink and necessary modification were done by adding the custom Simulink blocks developed for the current use case.

5.2.2 Design Variables

In order to tune the function for the reference maneuver, four input parameters or design variables were selected (see Figure 5.7). As per the terminology used in CAMEO these tunable input parameters will be referred to as the **variation parameters**. The variation parameters selected for the tuning task are:

- Acceleration Maximum (**A_MAX**) limits the maximum positive acceleration the vehicle can have while safely completing the maneuver. The negative acceleration is not limited in order for the vehicle to generate the necessary breaking force in case of obstacles.
- Jerk Maximum (**J_MAX**) limits the maximum positive jerk request from the controller function in order to meet the reference velocity curve. But only the positive jerk given by the engine and responsible for positive acceleration is limited, while there is no lower limit for the negative jerks for reasons mentioned previously.

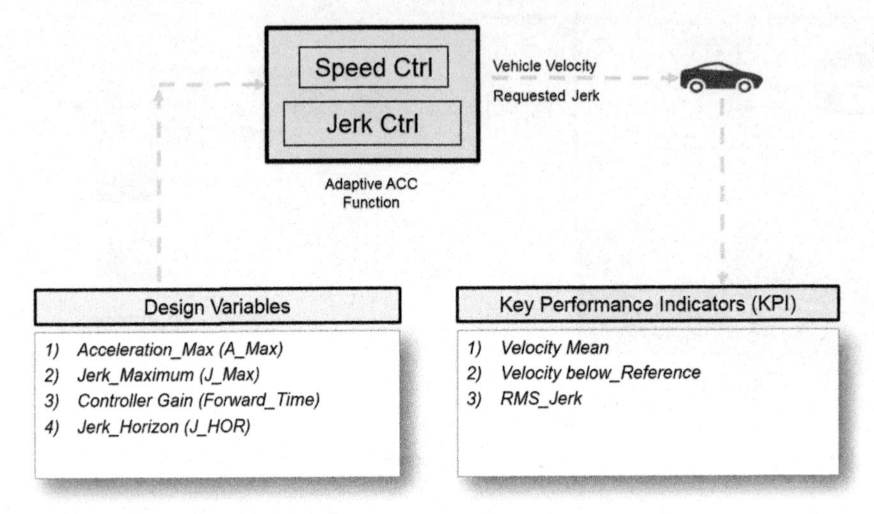

Figure 5.7 Function overview.

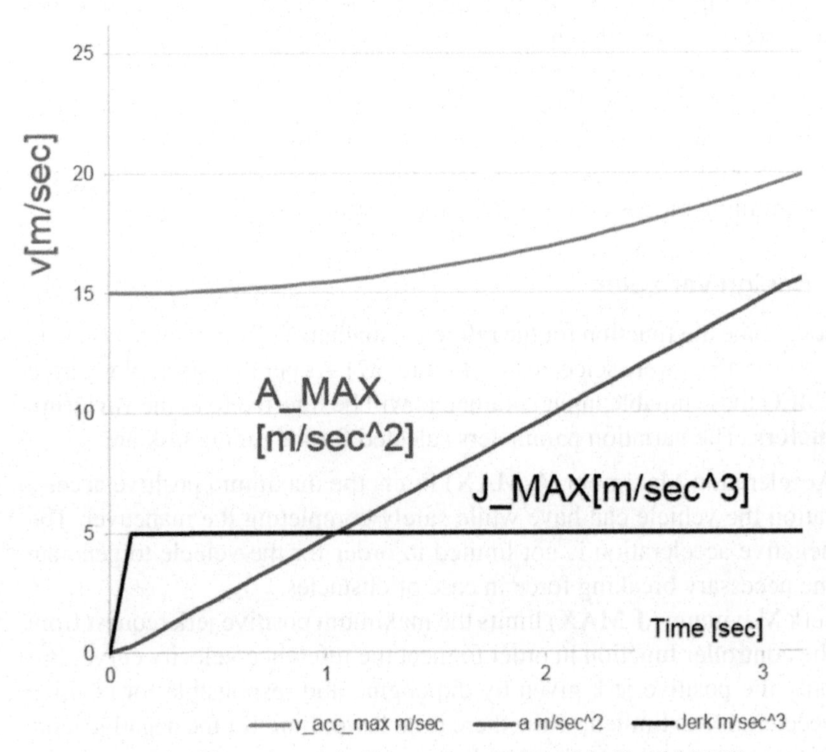

Figure 5.8 Illustration of the kinematic variables A_MAX and J_MAX.

Figure 5.8 illustrates the kinematic parameters, with acceleration being the derivative of velocity and jerk the derivative of acceleration.

- Forward Time (**FORWARD_TIME**) is a gain factor to transform the jerk request from the controller function to an acceleration request. Even though the controller function is based on jerk and sends the desired jerk requests for the vehicle, the interface to control vehicle motion is based on acceleration. Hence to control the vehicle the desired value of acceleration is required. In order to obtain the desired acceleration from the request jerk, one has to look forward for a given time which is called Forward Time. Mathematically it can be defined by the formula.

  A_req = A_0 + J_req*__FORWARD_TIME__
  A_req is the Acceleration request
  A_0 is the current vehicle acceleration
  J_req is the Jerk request generated by the controller function

- Jerk Horizon (**J_HOR**) is a parameter used to determine when the controller function sends the necessary jerk requests and the required jerk magnitude in response to an approaching curve. To define what is "near" and "far" (with respect to the distance from the approaching curve) for the controller function, the parameter **J_HOR** is used, where HOR stands for the horizon points (of the electronic horizon) to be considered. J_HOR is always a negative value, and values closer to zero make the controller respond to the approaching curve when it is further away with a smaller deceleration demand. Higher negative value tells the controller to respond when the approaching curve is closer in proximity but with a larger deceleration. A pictorial representation is given in Figure 5.9.

  The black line represents the target velocity set for the controller and the reference velocity curve is given in red. As explained previously the controller tries to control the vehicle speed (in blue) as close as possible to the reference speed.

  The mathematical expression "**A_MAX** + **J_HOR***time" determines the funnel of the vehicle velocity curve shape (shown in blue). More negative **J_HOR** give the velocity curve a sharper shape, while values closer to zero give the velocity curve a flatter shape.

The range of the variation parameters examined in the tuning task have been shown in Table 5.1.

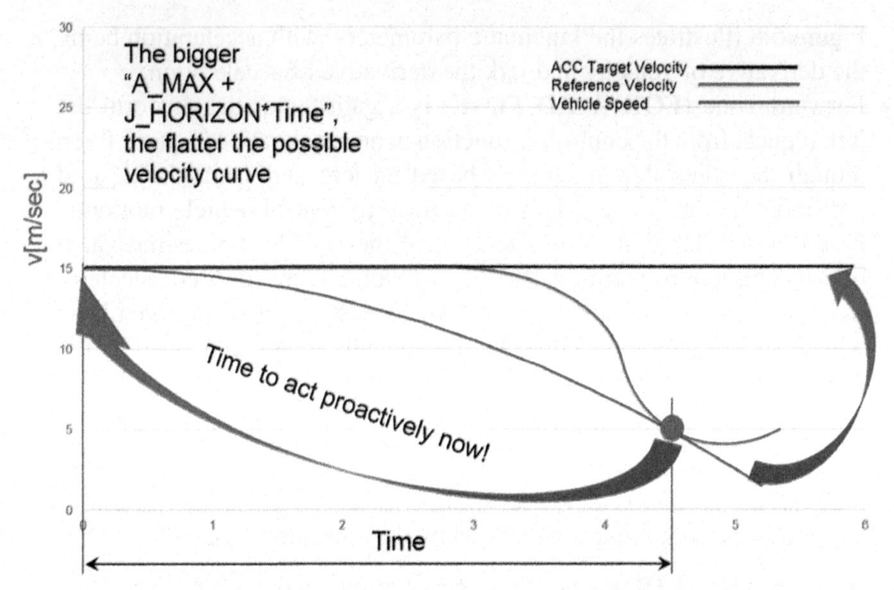

Figure 5.9 Illustration of the design variable (variation) J_HOR.

Table 5.1 Range of variation parameters used in the tuning task

Design Variable	From	To
A_MAX (m/s^2)	1	5
FORWARD_TIME (s)	0.1	2
J_HOR (m/s^3)	–5	–0.2
J_MAX (m/s^3)	1	3

5.2.3 Key Performance Indicators (KPI)

The output variables to demonstrate the effectiveness of our tuning task to meet the targets are described below and illustrated in Figure 5.10:

- Mean Speed: The mean of the vehicle speed in each test run is indicative of the sportiness of the driving experience. A higher mean speed helps finish the test maneuver in less amount of time, and makes the driving experience sportier.
- Speed below reference: The reference speed curve is the maximum speed with which the vehicle (Fiat 500L) can maneuver the digital test track without leaving the road for the reference use case. Hence to ensure vehicle safety it was ensured that the vehicle speed during the tuning task was always below the reference velocity.

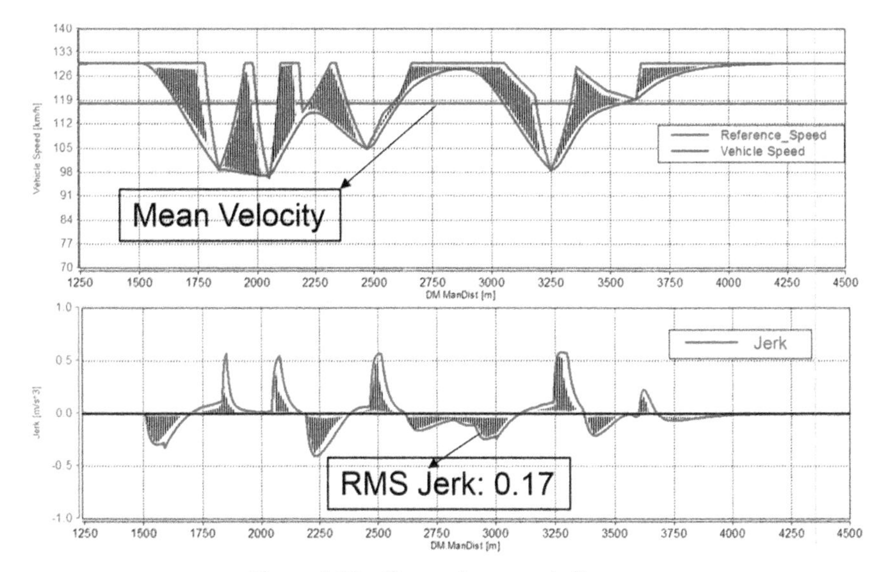

Figure 5.10 Key performance indicators.

- Jerk_RMS: Vehicle jerk which is the rate of change of vehicle acceleration, is indicative of the driving comfort. Lower rate of change of jerk gives a comfortable ride, so the root mean square of the jerk in a test run is a good indication of the driving comfort.

5.2.4 Test Maneuver

The test maneuver consisted of 5000 m test run on a digitized road imitating the road between Ceva and Savona in Italy run on IPG Carmaker for Simulink (CM4SL). IPG Carmaker environment is illustrated in Figure 5.11. The top left is the Carmaker for Simulink main GUI, showing details about the vehicle, simulation speed, time and distance of maneuver etc. The bottom left imitates the car instrumentation. The top right is time based plot of car speed and the vehicle jerk. The bottom right is the IPG Movie which illustrates the overall test run in a movie.

5.2.5 Test Run Overview

The test run overview is illustrated in Figure 5.12. The test parametrization was done in AVL CAMEO, where a space filling DoE design with the four variations was used. The variations were then uploaded to CM4SL through

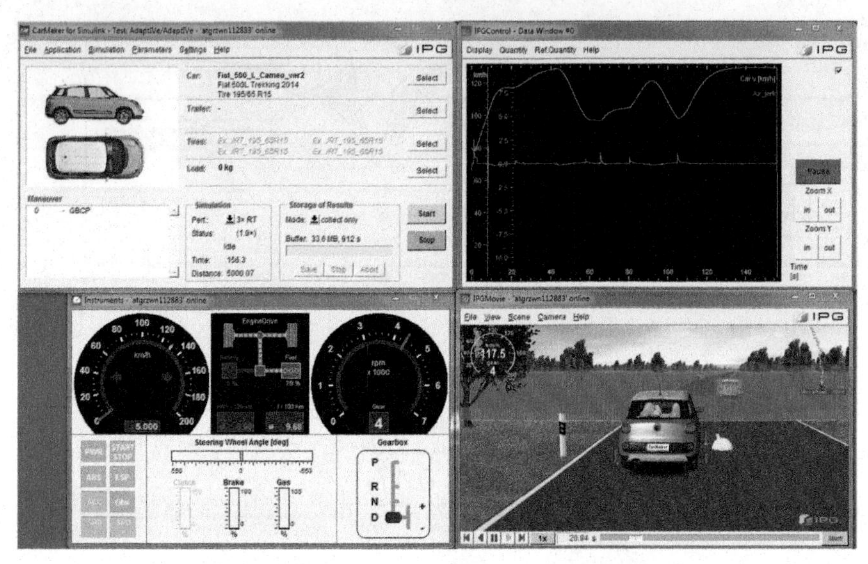

Figure 5.11 IPG Carmaker test environment.

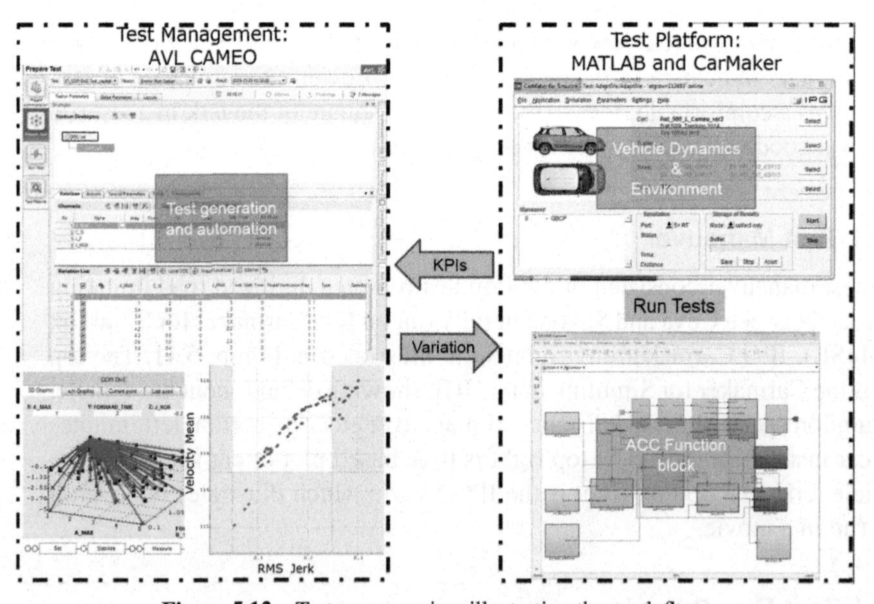

Figure 5.12 Test run overview illustrating the work flow.

the CAMEO-Carmaker Interface, where the test maneuver was run for each variations setting. AVL CAMEO then stores the measurement parameters observed as the KPIs for further evaluation.

During parametrization there were limits set on the minimum (–2 m/s^3) and maximum (2 m/s^3) acceptable vehicle jerk values. Whenever the vehicle jerk value violated the limits the test run at that test point was halted and no measurements were recorded. This affected the overall DoE design effectiveness with a reduced design space and as a result reduced measurement points. To overcome this challenge a COR DoE (Customized Output Range) method was utilized, which is an iterative method where first alternate test points were added by CAMEO to maintain the DoE design. Then based on these preliminary measurements the design space was further modified and additional test points were added in the relevant variation space to improve the final information from the measurements. Design space modification. The AVL CAMEO interface is illustrated in Figure 5.13, where the image to the left illustrates the overall test parametrization while the image to the right shows the test run window.

5.2.6 Raw Data Plausibility Check

Before the mathematical modeling of the selected output measured variables, the raw measurements were checked for plausibility. Firstly, the measured variables were checked for any outliers as shown in Figure 5.14 for mean

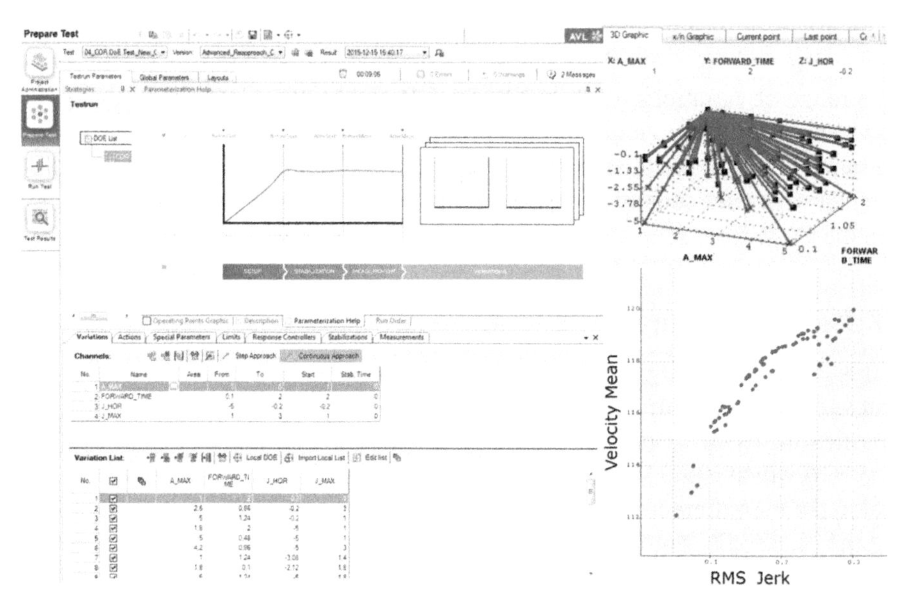

Figure 5.13 Left image illustrates the test preparation window while the right image illustrates the test run window.

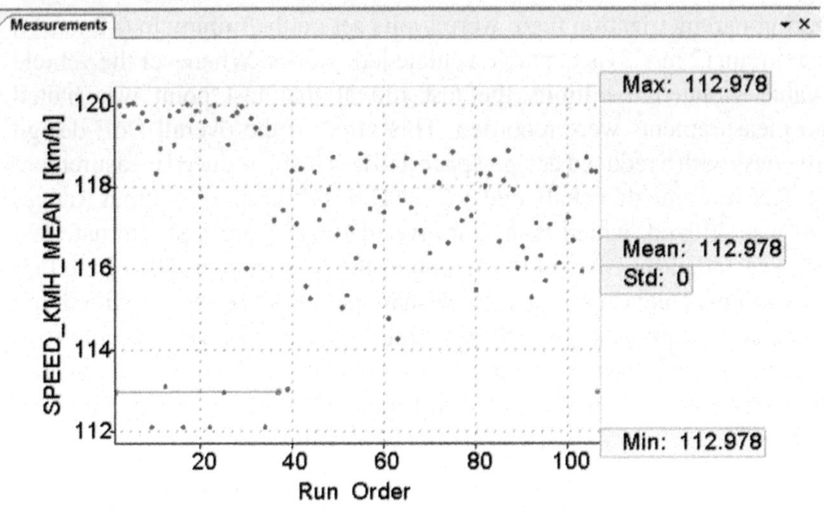

Figure 5.14 Checking for outliers in the measured variables.

speed. The measured values were within the acceptable range. The figure also shows that the repetition points (a select number of test conditions, usually the start condition which are repeated to check the reproducibility of test results) shown in green were perfectly reproduced.

The effect of design space modification, due to limit violations and the design correction by COR DoE method can be seen in Figure 5.15. In a certain range of variations for A_MAX, J_HOR and FORWARD_TIME there are no test points. Limit violations encountered when tests were carried out at these range of points are the reason why they were skipped by AVL CAMEO. Conversely a greater density of test points in certain ranges of variations show where the COR DoE added alternate or additional test points.

5.2.7 Meta Modelling

The raw data plausibility check was followed by empirical modeling of the output variables. The automatic modeling in CAMEO gave reasonable results with a neural networks model with local model order 2, as can be seen in Figure 5.16 which is the Measured (Predicted) plot which shows the fit of the model to the measurement points. If there is a perfect match all points will lie along the black line, but in our case the measurement points are reasonably close to the black line.

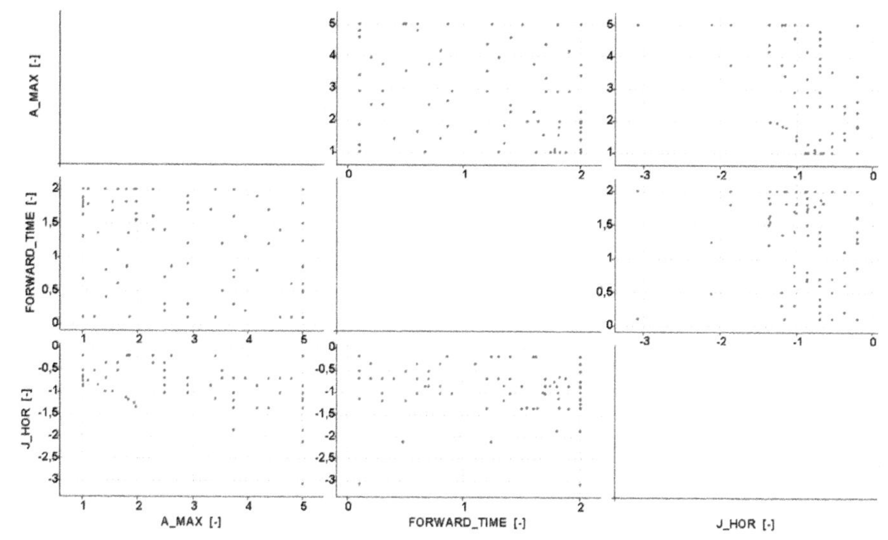

Figure 5.15 Check of DoE design and the boundaries of variation parameters.

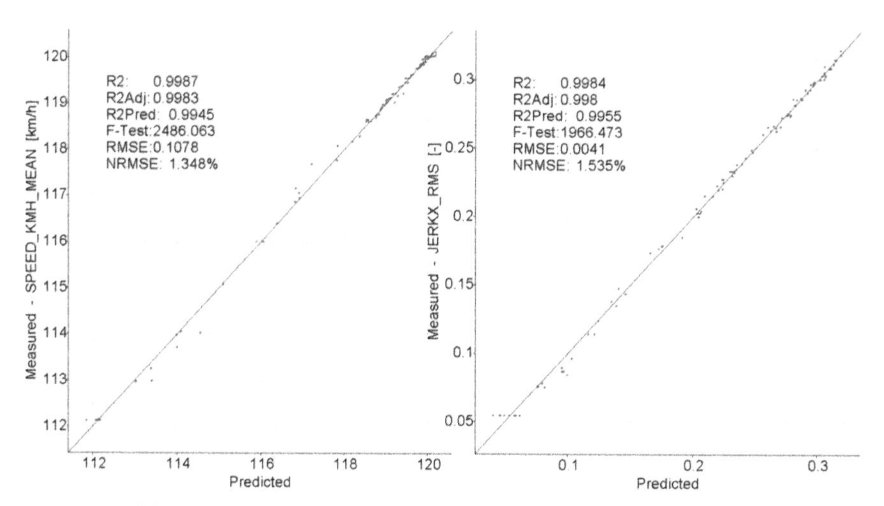

Figure 5.16 Figure depicting the quality of empirical modeling.

After checking the quality of modeling, the intersection plots were used which represent a cut through the multidimensional model, showing the influence of each variation depending on the values of the other variations. In Figure 5.17 the influence of the variation parameters on Speed_Mean and

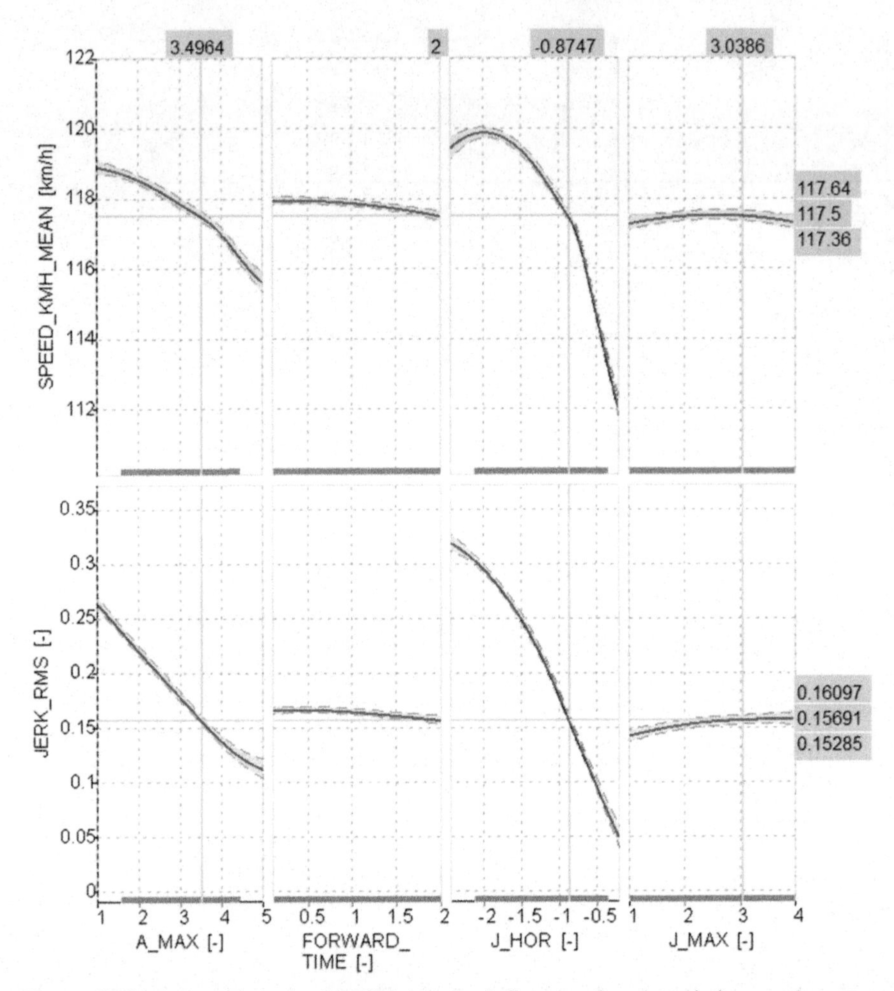

Figure 5.17 Intersection plot highlighting the influence of each variation on the output variables and their interaction.

Jerk_RMS can be observed. The confidence interval of the model is displayed in the green dotted line and colored section. The narrow confidence interval shows a high quality fit. The green bar on the x axis for each variation shows the total design space, and as the confidence interval of the model in the extrapolated region is also narrow, it shows good extrapolation capability of the model. Now looking at the intersection plots, it can be noticed that J_HOR and A_MAX have a strong influence on the output parameters. The more negative the J_HOR, the later the vehicle reacts to an approaching curve. Hence it is still travelling at a high speed before decelerating to approach the curve safely. Hence a higher mean speed is observed, but the resulting braking produces higher vehicle jerk reducing the driving comfort. Influence of A_MAX can be a bit counter intuitive but it can be seen that A_MAX is used to calculate J_HOR. The higher A_MAX, the less negative is J_HOR. Hence for higher A_MAX values J_HOR is closer to zero hence a smoother and slower ride. It can also be observed that higher FORWARD_TIME allows for a smoother and slower ride, which is because the controller can take more time to achieve the desired acceleration.

5.2.8 Optimization

From the intersection plot, it is possible to manually find values of the variations which give a comfortable ride or sporty ride or an acceptable compromise. But it is quite easy to miss the optimum or an acceptable compromise when working with multiple input variations, hence the optimization tool in CAMEO was used. In the current tuning scenario, the target was to be able to isolate two modes of operation, comfort mode and sporty mode. Hence a multi objective optimization was chosen with limits set on the minimum desired mean speed of 115 Km/h and maximum acceptable JERK_RMS of 0.28 (Figure 5.18).

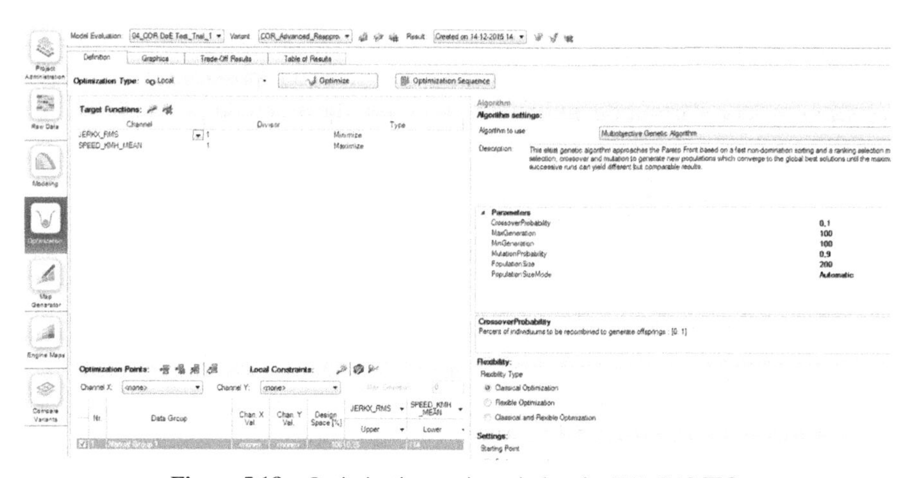

Figure 5.18 Optimization setting window in AVL CAMEO.

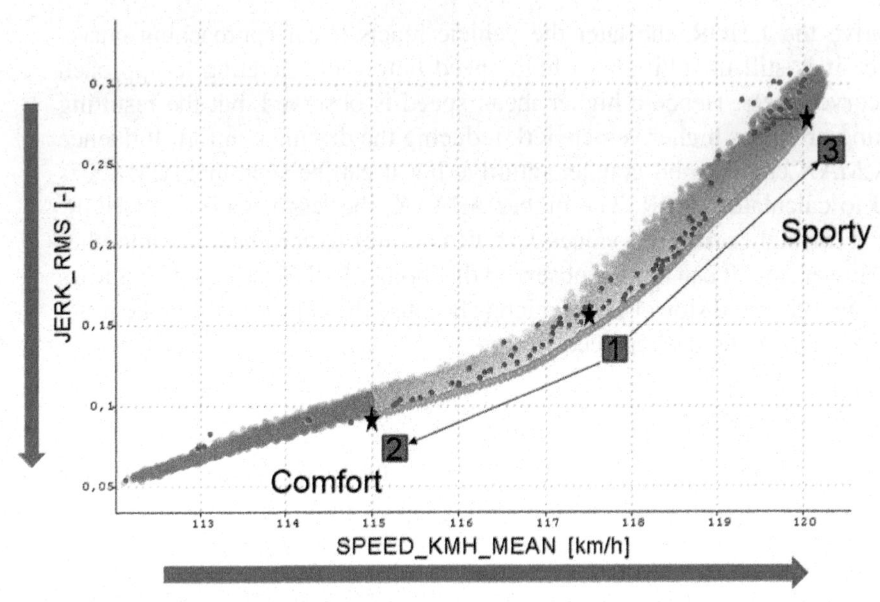

Figure 5.19 Trade-off plot between comfort and speed.

The result is plotted in a trade-off plot as shown in Figure 5.19, where the steel blue is the pareto front, the blue points indicates the measurement values and the other yellows points are random space filling points. The pareto front shows the possible optimum trade-off solutions which can be considered equally good as the only way to improve on objective would be to compromise on the second objective. So by observing the pareto front it is possible to define an optimum for comfort mode and an optimum for sporty mode of operation Table 5.2.

In Figure 5.20: Sporty mode vs comfort mode: the vehicle performance when operating at the two modes can be observed. The red velocity curve is the reference velocity and blue velocity curve is the actual vehicle velocity. It can be observed that the actual velocity is always below reference velocity which was the safety requirement. Also the velocity changes in comfort mode

Table 5.2 Variations values for comfort and sporty mode

	A_MAX	FORWARD_TIME	J_HOR	J_MAX	SPEED_Mean	JERK_RMS
Comfort	4.99	1.94	−0.84	1.0	115	0.09
Sporty	3.88	1.37	−1.84	3.36	120	0.28

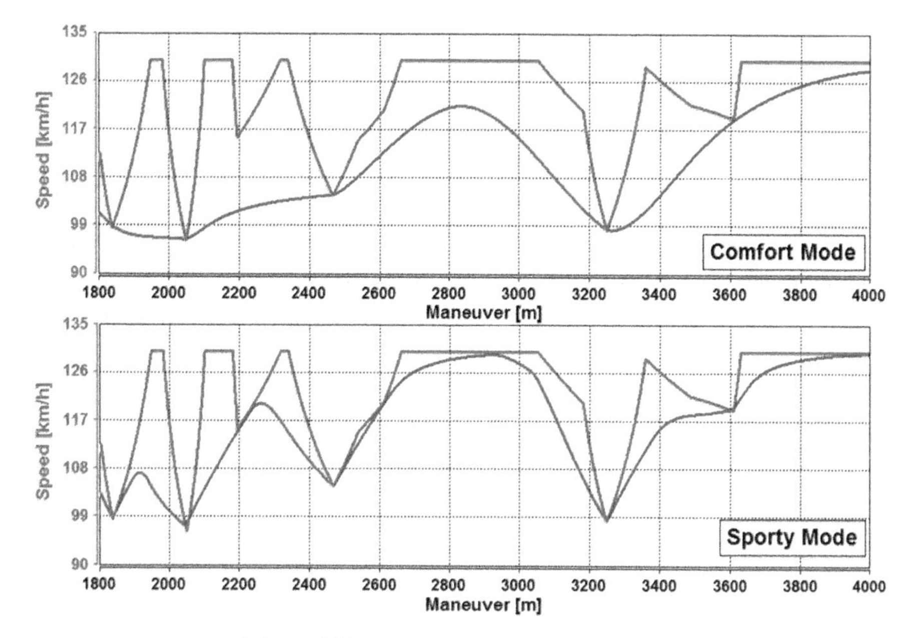

Figure 5.20 Sporty mode vs comfort mode.

is more gradual with no sharp peaks unlike in sporty mode where there are rapid fluctuations in vehicle velocity. This behavior is also mirrored in the acceleration values in both operation modes. The vehicle jerk curves (red plot is the jerk request generated by the controller and blue the actual vehicle jerk response) show much lower values in vehicle jerk for comfort mode while the sporty mode show sharp and frequent peaks in jerk value.

5.2.9 Verification

The pareto front consists of points a majority of which are from the model extrapolation. In order to verify the robustness of the model to accurately extrapolate, ten random points were selected from the pareto front and for the corresponding variation values the test runs were rerun. The results from these test runs were evaluated as verification points in CAMEO. The Figure 5.21 shows the extrapolated model (in red) and its prediction interval (in blue), and the measured verification points and its modeling (in green). The measured verification points lie within the prediction interval of the model, showing the extrapolation accuracy of the model.

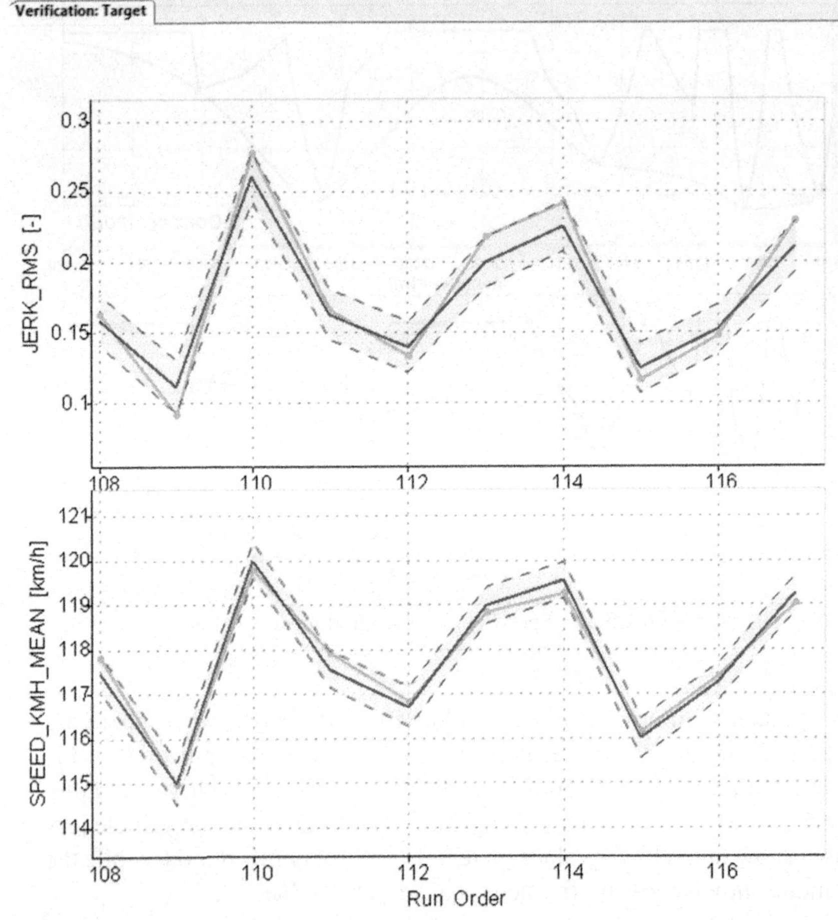

Figure 5.21 Verification plot to see how well the measured results from the verification run fit the model results.

5.3 Model-based Validation

Once the reference tuning task is completed, it has to be tested, if the tuning results are still acceptable, when not running the reference use case but for varying road characteristics. Will the comfort mode still allow for a comfortable drive also for different road situations? It would be unfeasible to run simulations on thousands of different roads, besides making it difficult to realize the influence of a specific road. In the current method the two tuning modes are fixed and a system variation of a digitized road is performed using the model based approach to validate our tuning results.

The digitized road is shown in Figure 5.22, where the lengths of the straight sections (L1, L2, L3, and L4) and curvatures (R1, R2, R3) were varied while keeping the total maneuver length to 5000 m. The controller settings were fixed to run at first comfort mode and then sporty mode, and the resulting measurement output variables are shown in Figure 5.23.

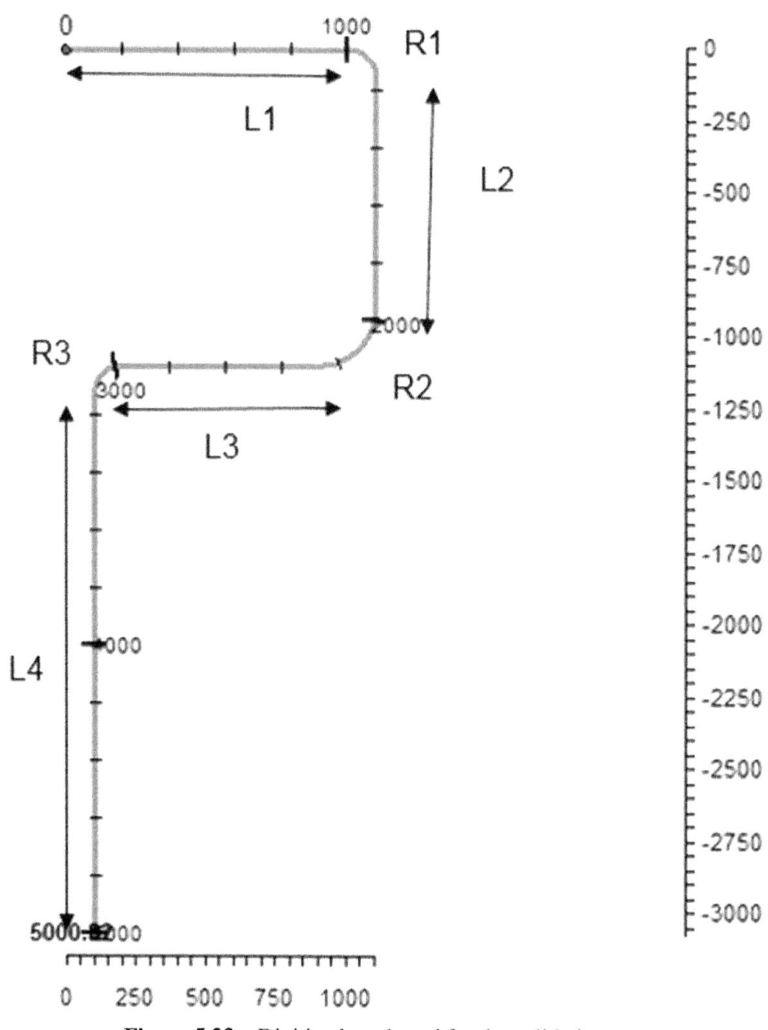

Figure 5.22 Digitized road used for the validation run.

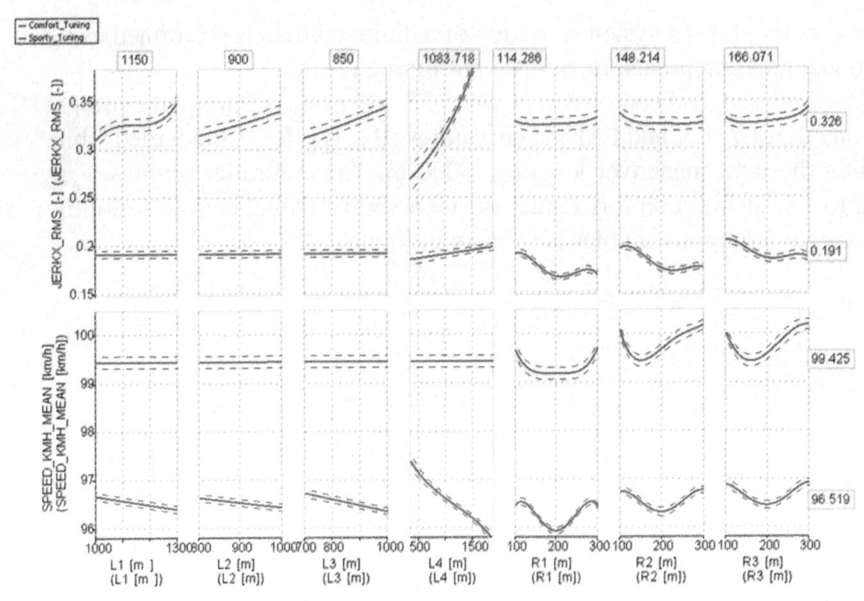

Figure 5.23 Measurements comparison when run on comfort mode (in blue) and sporty mode (in red).

It can be seen in Figure 5.23: Measurements comparison when run on comfort mode (in blue) and sporty Mode (in red) that for the sporty mode the resulting drive comfort is lower as indicated by the higher JERK_RMS. The length of the straight portions do not influence the JERK_RMS for comfort mode as strongly as in the sporty mode. The curvature of the turns seem to influence the output in both the operation modes. A JERK_RMS limit of at least 0.35 is expected, and it can be seen that the limit is maintained in both the modes of operation for majority of the design space. In the sporty mode the controller is set to maintain a higher vehicle speed and responds to the oncoming curve only when it is close, hence the longer the straight sections, the larger the jerk experienced when it decelerates rapidly to approach the curve followed by a strong acceleration on leaving the curve. For the comfort mode, the controller is set to focus on keeping the vehicle jerk close to minimum. The validation task showed that, if the function (our UUT) is kept constant and the simulation environment is changed, the function still manages to meet the expected vehicle jerk targets. The influence of 'L4' on the jerk behavior needs to be further investigated as it strongly increases the vehicle jerk fluctuations at higher values especially for the sporty mode. To further explore and investigate the influence of test track characteristics on the function response, it can be tested on a variety of road types and test tracks. This assists in the further improving the function performance.

5.4 Conclusions

Virtual tuning of an ADAS function developed on a MiL environment using an optimization tool can be a powerful combination for the development of a brands driver assistance system. The classical approach relies on a subjective tuning of the ADAS function on a proving ground and public roads, which can be supported and accelerated by using a virtual tuning environment. Using DoE methods supported by AVL CAMEO, it was possible to increase the number of tuning tests compared to a manual tuning, and also the number of target parameters and tests needed to match them. The possibility to use the developed function for alternate use cases by separating the software and the tuning data is precondition for tuning works in general.

Independent of that also in the validation process a model-based approach can be very helpful, as the test coverage for a certain use case can be extended to a wide range of possibly occurring variants of that use case. The robustness of the key performance indicators considered as relevant can be estimated.

Acknowledgement

We would like to thank Mr. Andreas Saroldi from CRF for providing the ADAS function.

References

[1] M. Paulweber and K. Lebert: *Instrumentation and Test Systems: Power Train development, Hybridization and Electrification.* Chapter 5.4.2 *Application Data management,* Springer View, 2016.

[2] AVL Tuning Data Management Software: http://www.avl.com/creta (called at 2016 01 20).

[3] H.-M. Koegeler; A. Fürhapter; M. Mayer; K. Gschweitl: *DGI-Engine Calibration, Using New Methodology with CAMEO.* In SAE NA, Capri –Italy, 23–27. September 2001.

[4] E. Castagna; M. Biondo; J. Cottrell; H. Altenstrasser; Ch. Beidl; H.-M. Koegeler; N. Schuch: *Multiple Tier 3 Engine Applications based on global modelling.* In MTZ 6/2007.

[5] T. Fortuna; H.-M. Koegeler; M. Kordon; V. Gianluca: *DoE and Beyond-Evolution of the Model based Development Approach,* in ATZ worldwide, Springer, 2015.

[6] H. M. Koegeler; B. Schick; P. E. Pfeffer; A. Contini; M. Lugert; T. Schöning: *Model Based Steering ECU Calibration on a Steering in the Loop Test Bench,* in Chassis Tech 2015.

[7] D. C. Montgomery: *Design and Analysis of Experiments,* John Wiley and Sons.

[8] A. Rainer; H. M Koegeler; D. Rogers: *Iterative DoE – Improved emission models and better optimization results within a shortened measurement time,* in PMC, 2014.

PART II

Test Case Functions

6

Deep Learning for Advanced Driver Assistance Systems

Florian Giesemann[1], Guillermo Payá-Vayá[1], Holger Blume[1], Matthias Limmer[2] and Werner R. Ritter[2]

[1]Institute of Microelectronic Systems, Leibniz Universität Hannover, Hannover, Germany
[2]Vision Enhancement, Daimler AG, Germany

6.1 Introduction

Today, vehicles contain a wide range of electronic driver assistance systems. These systems, for example *Anti-lock Braking System* (ABS) or *Electronic Stability Control* (ESC), increase car safety and on a more general level even road safety. More complex *Advanced Driver Assistance Systems* (ADAS), like *Lane Departure Warning*, *Overtaking Assistant*, *Collision Warning* or *Emergency Breaking* do not only observe the parameters of the vehicle itself, but also require information regarding the environment. Future applications, which target autonomous driving, need an even more detailed understanding of the vehicle's environment and the current driving situation. Therefore, vehicles are equipped with a number of sensors, which enable the perception of the vehicle's surroundings including other road users. But the sensors generaly used deliver a huge amount of raw and unrefined data, from which the necessary information needs to be extracted. For instance, for camera sensors, an algorithm called *Scene Labeling* can be used to detect relevant objects in camera images. It assigns every pixel of an input image to a semantic class (e.g., road, car, free space etc.) and can therefore be used to extract detailed information from the scene.

The increasing complexity of algorithms and the increasing amount of data that has to be processed requires a high amount of processing power. At the same time, processing hardware is subject to restrictions regarding power consumption and size. These conditions make the field of embedded hardware platforms for driver assistance systems challenging.

This chapter is organized as follows: Section 6.2 gives an introduction to Scene Labeling techniques and their application in Advanced Driver Assistance Systems. Section 6.3 explains the concepts of Convolutional Neural Networks and Deep Learning. In Section 6.4, an exemplary CNN is presented and evaluated. Section 6.5 describes different hardware platforms for Scene Labeling. Finally, Section 6.6 summarizes the chapter.

6.2 Scene Labeling in Advanced Driver Assistance Systems

Getting a thorough understanding of the vehicle's environment is an important step in the development of advanced driver assistance systems. Different techniques for detection and classification of objects have been developed. Literature offers a wide range of algorithms for detecting traffic signs, traffic lights, driving lanes, and also other vehicles and pedestrians. In order to build up a comprehensive understanding of the environment, not only single objects have to be detected, but also the objects in relation to each other have to be determined. This is commonly referred to as *Scene Labeling*.

Scene Labeling is a technique to classify images on different levels of detail. Image-level Scene Labeling (e.g., [1]) is used to derive one or more labels for the whole image that describe different scene types, e.g., urban, inter-urban, or highway. On another level, labels are deduced for small sub regions of an image, so called *regions of interest*. This allows for a more detailed understanding of the scene in terms of objects, like pedestrians, vehicles, driving lanes, traffic signs and so on. On a third level of detail, each pixel in an input image is classified and provided with a semantic label. The information provided by these labels can be used in different applications, for example in pedestrian/obstacle detection, close range lane course estimation or relative map positioning.

Scene Labeling can also be combined with other detection methods in order to increase reliability and thereby increase the integrity level of safety functions. Moreover, it can replace different detection modules in order to save resources.

The Scene Labeling task is usually performed in two steps. The first step extracts features from the input image; the second step computes a classification of the image, the region, or the pixels from the extracted features.

Several different features are used in order to perform image segmentation and semantic labeling. Some algorithms rely on single, low-level features, like color [2], texture [3, 4], shape [3, 5], geometry [6], and edge features [7]. Object detection algorithms are used to extract high-level features, e.g., pedestrian detection [8], traffic sign detection [9], and lane detection [10]. Some algorithms perform labeling using image segmentation techniques, e.g., Super Pixels [11] or sliding windows using Boosting [12] to detect regions of one certain class, e.g., pedestrians or traffic signs.

Classification of extracted features is performed using different techniques, like Support Vector Machines [13], Genetic Algorithms [7], or Neural Networks [14]. Probabilistic models like Conditional Random Fields (CRF) [15] and graph-based optimization methods (e.g., Graph Cut [16]) are used to combine different features and include smoothness constraints or neighbor relationships.

Recent advances in the field of deep learning and neural networks yielded a new technique for the scene labeling problem, which is described in the next section.

6.3 Convolutional Neural Networks and Deep Learning

Typical systems for detection and recognition of objects or situations use a two-step data processing scheme. In a first step, features are computed from data gathered through different sensors, like cameras, radar, etc. Then, a second step uses the previously computed features in order to classify the candidates into the object classes. The implementation of the classification step might involve the use of machine learning techniques, i.e., the training of a classifier. One difficulty in this scenario is the selection of features to be used. Often, these features are hand-crafted and a lot of work might be involved in tuning the parameters in order to find a set of features that can be used for reliable detection and recognition of objects.

Another way of building recognition systems that evolved recently is the use of learning techniques and especially the technique of *deep learning* with close coupling between the feature extraction and feature classification steps. Deep learning describes methods, in which feature extractors are not hand-crafted but automatically learned from a set of training data. Multiple layers of feature extractors can be used in a hierarchical structure in order to allow deeper layers to extract features of higher order from previous layers. The idea behind this technique is that the learning algorithm is capable of detecting the best features for the following classification step itself. Commonly used implementations of the deep learning methodology are *artificial neural networks*.

6.3.1 Introduction to Neural Networks

Inspired by processes in the biological neural networks of the central nervous systems and especially the brain, different computational models of artificial neural networks have been developed [17]. Artificial neural networks are built as a collection of relatively simple units, so called neurons, that are connected together to form a network which can process a complicated task. One of the first models of neural networks is called *perceptron* [18]. The simple perceptron neurons perform binary decisions depending on their input values. The input signals x_i are weighted and accumulated. The neuron "fires", i.e., produces an output signal y of 1, if the weighted sum of the input signal exceeds a given threshold value, and outputs 0 otherwise. The first networks had one single layer of neurons and were only capable of computing linear classifications. More complex networks with multiple layers were capable of computing more complex classifications. Nowadays, neural networks use a different model for the artificial neurons [19, 20], as depicted in Figure 6.1. The input values, which are now real numbered values, are weighted and accumulated. Afterwards, a non-linear activation function is applied to the sum. Commonly used activation functions are the *sigmoid function*, which can be interpreted as a smoothed threshold. Recently, *rectifier linear units (ReLU)* have been reported to have several advantages over the sigmoid functions [21]. Some exemplary activation functions are shown in Figure 6.2.

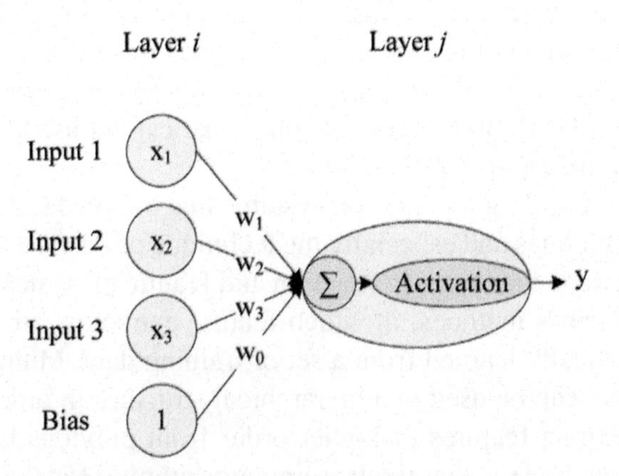

Figure 6.1 Model of an artificial neuron.

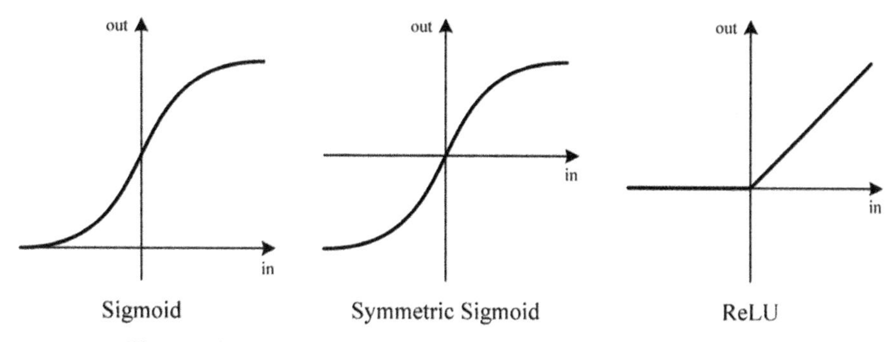

Figure 6.2 Exemplary activation functions used in neural networks.

The bias is another value summed up along with the weighted inputs. This parameter influences the neuron's general activity or the likelihood for an output activation of the neuron. For simplicity, the bias can be interpreted as the weight for a constant input value of 1, so that all parameters of the network can be interpreted as weights. Therefore, a neuron with inputs $x_1, x_2 \ldots, x_n$, weights $w_1, \ldots, w_n$, bias w_0, (with $x_0 = 1$) and activation function f can be described mathematically as

$$y = f\left(\sum_{i=0}^{n} w_i x_i\right).$$

In so called *Multi Layer Perceptrons* (MLP), neurons are arranged in layers. The neurons of one layer are connected to neurons in the following layers. No connections exist between neurons of one layer and the graph formed by the neurons and connections is a directed acyclic graph. Therefore, MLPs are called *feed forward networks*.

The task performed by the neural network depends on the parameters, namely the weights and biases. Therefore, the network parameters have to be adjusted before the network produces the correct outputs. This adjustment is called *training*. Different methods for training multi-layer feed-forward networks have been devised. The most commonly used technique is the backpropagation of error [22].

6.3.2 Supervised Learning

In a neural network, the internal parameters (weights of the neurons) are also called *trainable parameters*, since they can be trained to approximate a desired function. In case of Scene Labeling, this function would map a pixel of an image to a specific label, using the pixel's neighborhood. For classification tasks with a given set of classes, *supervised learning* schemes are used. A set of *training samples* contains input images together with the desired output. In combination with an error function, the training set can be used to adjust the internal parameters of the network.

The Cost Function and Backpropagation

Supervised learning for neural networks is performed by measuring the neural net's estimated output against the expected output with a so called *cost function*. The goal of a supervised training is to find the internal parameters which minimize this cost function regarding a set of training examples. Since the network in general models a highly non-linear function, *gradient descent* can be used as an optimization procedure. This is done by computing the gradient of the cost function and leveraging the chain rule to propagate the cost and the gradient back through each layer of the network. The weights in each layer are updated according to the current gradient of the backpropagated cost. This algorithm is therefore called *backpropagation*.

A successful training converges against the minimum value of the cost function. It is important to choose the cost function suitable for the task that the neural network needs to perform. For classification tasks, a combination of the softmax function and (multinomial) logistic regression is often performed to train the internal parameters. The softmax function serves as a normalization function, which maps input values x_j of arbitrary range to values in the range $(0, 1)$ that add up to 1. The maximum of the input values maps close to 1 while the other values map close to 0. The function is defined by

$$\text{softmax}(x_j) = \frac{e^{x_j}}{\sum_{k=1}^{K} e^{x_k}} \quad \text{for } j = 1, \ldots, K.$$

The softmax directly serves as the multinomial version of the logistic function used in logistic regression. The resulting cost function is defined by

$$\text{cost}(x) = -\ln(\text{softmax}(x'_k)),$$

with x'_k as the predicted output of the neural network for the actual class k. The cost is therefore the negative log-likelihood of the expected class, which minimizes, when the estimated probability for that class is 1.

Stochastic Gradient Descent

Gradient descent is an algorithm that finds a local minimum by following iteratively the negative gradient of a function $F(x)$ at each point x. It can be defined as

$$x_{i+1} = x_i - \eta_i \nabla F(x_i).$$

Here, η_i is the so called *learning rate* at iteration i. Choosing the right η in every iteration of the algorithm is crucial for the success and the convergence speed of the optimization. If η is too small, it takes many iterations to find a local minimum. Furthermore, the detected local minimum might just be a plateau with better local minima in the neighborhood. If the chosen learning rate is too big, it is possible to jump repeatedly over the local minimum, but never reaching it. In severe cases, it is even possible that the algorithm diverges. There are several schemes for choosing the learning rate adaptively. Resulting in most cases in a computational overhead, which is due to an additional analysis step at the current point of the function. A fixed learning rate is often used, which is scaled down in every iteration. Later iterations are supposed to be close to a minimum and require therefore a finer grained learning rate.

Given the basic gradient descent update rule, the term $\eta_i \nabla F(x_i)$ can be called update ν_i of iteration i. Since these updates only rely on the current gradient, small bumps in the error function might lead to a jittering path in the gradient descent, which increases the number of iterations until a local minimum is found. This might especially occur in stochastic gradient descent, which does not use every training sample in each iteration. To overcome this, many learning schemes extend the update rule by a *momentum term.* The update rule is then defined by

$$x_{i+1} = x_i - (\eta_i \nabla F(x_i) + \mu \nu_{i-1})$$

with a new definition for the update ν_i:

$$\nu_i = \eta_i \nabla F(x_i) + \mu \nu_{i-1} \quad \text{and} \quad \nu_0 = 0.$$

The parameter $\mu \in \mathbb{R}(\mu \geq 0)$ denotes the influence of the update from the previous iteration. If $\mu = 0$, no momentum is used to calculate the current update. Update steps are stabilized and the "velocity" in flat valleys of the error function is increased by using a momentum. However, this property is not always desired in all gradient descent schemes, because the momentum might also cause the update to overshoot. Hence, the momentum term should be used with care.

In a learning environment, a point x of the cost function is the set of internal parameters unified with the expected net output. Since there is not only one training example but many, there are also many expected output points. The cost of more than one data point is therefore the sum of all costs.

This is called *objective function*. It follows, that in an iteration (epoch) of the gradient descent algorithm, all data points need to be processed. This is called *batch gradient descent*. In many cases though, processing all data points in one epoch is not feasible because of the size of the dataset. In this case, *stochastic gradient descent* is used. Instead of predicting all data points per epoch, a random subset for each epoch is generated. If the subsampling is random enough in each epoch, this method optimizes an approximation of the objective function. Though each individual epoch might not sufficiently approximate the objective function, the repeated random sampling does. Stochastic gradient descent is therefore a common approach to train a neural network with big datasets.

6.3.3 Convolutional Neural Networks

A *Convolutional Neural Network* (CNN) is an extension to the common MLP, originally designed for two-dimensional data, like images. As the name suggests, it adds *convolutional layers* to the set of possible layers in an MLP. There is an analogy here with the primary visual cortex of a cat, which also uses convolution-like simple cells to extract information from spatially close overlapping regions of the field of view [23]. In [24], the authors showed that the backpropagation algorithm can be extended for the training of CNNs by introducing an update and backpropagation rule for convolutional layers.

Convolutional Layer

The convolution layer differs in two ways from the common fully connected layer of an MLP:

1. Convolution layers only sum up a fixed window of the input signal. They are therefore only *locally connected*. This connection window is called *receptive field* of the layer.
2. Each possible position of a receptive field uses the same weights to produce an output. This is called *weight sharing*.

The output signal is produced in a sliding window fashion, by applying a weighted summation of the receptive field for each possible receptive field position. The output contains as many values as possible positions. It is exactly a convolution of the input signal, where the layer weights form the convolution filter (kernel). A convolution layer can have several filters, thus forming a *filter bank*, which is analogous to the amount of hidden units in this layer.

Pooling Layer

Another important extension of the MLP is the *pooling layer*. A pooling layer performs a subsampling of the input signal, by "combining" small windows of the input signal into several singular values. A common pooling function is *max-pooling*, which calculates the maximum of its receptive field. Another pooling function is *average-pooling* which computes the average value in its receptive field. A pooling can be seen as a convolution with a special function and a *stride* that equals the filter size of the *pooling kernel*. Regular convolutions have a stride of 1, meaning every pixel position is computed in the convolution. A stride of 2 means that every other pixel position is computed. The purpose of pooling is not only to reduce the spatial size of the input signal, but also to increase the robustness of translational invariance of the activations.

Multiscale CNN

A variation of convolutional neural networks is the *Multiscale CNN*. Instead of processing an input signal as it is, the Multiscale CNN processes several scaled down versions of the signal simultaneously. This approach increases the ability to extract scale invariant features, without the need to increase the size for the extracted pixel neighborhood patch windows. The extracted feature maps of each scale are finally combined to produce a joint feature map. This can be done by a fully connected layer that takes all feature maps as an input to compute its output. For the Scene Labeling application, an image pyramid has to be created prior to the extraction of image patches for each scale, which are then fed to the Multiscale CNN.

Patch Based and Image Based Application

Neural networks for image classification tasks were traditionally designed so that they process a complete image of fixed size and produce classification results of a fixed size as well. Big image sizes automatically implied that the fully connected hidden layers had also a great amount of hidden units. This resulted in the reduction of the input images sizes to keep the neural networks scalable and computable. In order to apply neural networks in a pixel classification scheme, image patches had to be extracted at each pixel position that needs to be classified. In many cases, these extractions are applied sparsely across the image to produce a coarse pixel classification.

A patch based application of CNNs for pixel classification tasks is computationally very inefficient, because image patches for neighboring pixels overlap. Therefore, the same convolutions are computed multiple times.

This redundancy can be omitted by applying CNNs in an image-based fashion. This has an effect on several aforementioned components of the neural network, since they have been designed in regard to a patched based application. The fully connected layer especially is not applicable in an image based application, because *full connectivity* is contrary to the *local connectivity* of the convolution layers for arbitrary image sizes. The adequate translation of a fully connected layer in a patch based approach is actually another convolution layer, with a $1{\times}1$ convolution on all locally connected input values.

Another layer type that works differently in an image based application is the pooling layer. A naïve translation would result in a huge loss of output resolution, since pooling layers in patch based mode are designed to subsample the input signal. A patch based application on every possible pixel location though doesn't share this subsampling property. This is why the patched based approach really evaluates every pixel location, while an image based approach implicitly only fully evaluates a subset of all pixel location due to the subsampling. To remove the subsampling property, a pooling must be applied in a convolutional manner (overlapping pooling). Looking at the output maps of such an overlapping pooling, it is clear, that they differ from maps of a non-overlapping pooling. In particular, neighboring pixels from a non-overlapping pooling are not neighbors anymore. If a convolution layer follows, it results in a wrong calculation of the output maps. This can be corrected by reordering the pixels after the pooling layer into n subimages, where n is the size of the pooling kernel or the stride, and apply the following layers on each subimage independently [25]. The reordering is hence defined as *fragmentation*, because the input map is fragmented into smaller output maps. Figure 6.3 shows such a fragmentation after the application of a 2×2 pooling producing 2×2 subimages.

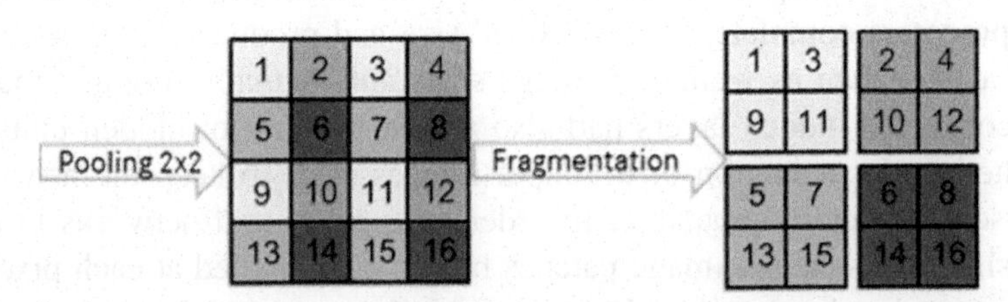

Figure 6.3 Example of a fragmentation after a 2×2 pooling. The naïve approach would only produce the bright pixels, while an overlapping pooling produces all other possible pixels (purple, green, and blue). These pixels must be reordered to be able to correctly continue with the forward propagation of the neural network.

For Multiscale CNNs, an image based application introduces another difficulty, which needs to be solved. In a patch based approach, the image patches for each scale have to be extracted and each patch has the same size. In an image based approach however, the feature maps for different scales are of different size. This becomes a challenge in the fully connected layer, which combines the feature maps of all scales. Since there are no fully connected layers in the image based approach, the feature maps of each scale need to be transformed so that a regular convolution layer can handle them. The simplest solution is to scale the smaller maps up so that they all match in size. If the maps have been fragmented because of a pooling layer, they need to be *defragmented* before they are scaled up. Defragmentation is the reverse function of fragmentation, turning multiple smaller maps into one bigger map.

6.4 CNN for Scene Labeling

There are many ways to perform Scene Labeling on images. CNNs have proven themselves useful on this task, because they achieve state of the art performance without the need to develop complex multi cue frameworks that combine different inputs and sensors. Additionally, many frameworks for modeling, training and execution of CNNs exist, e.g., Caffe [26], Torch7 [27], Theano [28], Pylearn2 which is built on top of Theano, and cuda-convnet [29]. These frameworks exploit the CNN's parallelizability to provide fast and time efficient implementations using *General Purpose GPUs* (GPGPU). Furthermore, the research community is actively training and publishing models, which can often be adapted to a specific task by resuming the training with corresponding data. Most frequently used models are AlexNet [30], GoogleNet [31] or VGG [32]. They differ in complexity and run time efficiency, but reached state of the art performance during their time of publishing for certain challenges on datasets like ImageNet [33]. A high *network capacity* is needed to achieve a high accuracy on such complex tasks. So the trained models are rather big and need a huge amount of computational power. Incorporating this into an embedded system with low power consumption, as is needed for ADAS, is still a great challenge.

The following section describes one possible model with reduced complexity, selected for implementation in the course of the DESERVE project. Its purpose is to detect the road, vehicles and vulnerable road users, which can then be utilized for lane prediction and pedestrian detection.

6.4.1 Exemplary Network for Scene Labeling

The proposed model is derived from the Multiscale CNN used in [34]. It consists of 2 convolutional layers and 2 pooling layers. The activation function, used after the convolutional layers, is the ReLU function (see Figure 6.2). Each convolution layer contains a bank of $16 \times (7 \times 7)$ filter kernels. These four layers are applied on three scales of the input image and combined by a fully connected layer, producing 6 output channels: *background*, *road*, *vehicle* (including cars, trucks, busses, ...), *vru* (vulnerable road users: pedestrians, cyclists, ...), *sky* and *infrastructure* (buildings, signs, barriers, traffic lights, ...). Those channels are normalized by a softmax layer to produce class probability maps for each class. By applying an argmax on these maps a class membership map is produced returning the most probable class for each pixel. The input images are preprocessed by transforming them into an image pyramid and locally normalizing them afterwards to zero mean unit variance in a 15×15 neighborhood. Figure 6.4 shows the complete toolchain and Figure 6.5 the network topology in more detail.

6.4.2 Evaluation

The topology described in subsection 6.4.1 was trained with 6895 labeled night time images of a near infrared camera used in the NV3 night vision system of a Mercedes Benz S-Class. The images show mainly rural, but also urban, road scenes under different weather conditions and different seasons. To augment the heavily under-represented *vru* class, 15174 images are added to the aforementioned set of images, where only the *pedestrian* and *cyclist* labels are used. This is called the *learn set*. The training scheme is stochastic gradient descent with the logistic regression objective function for 6 classes.

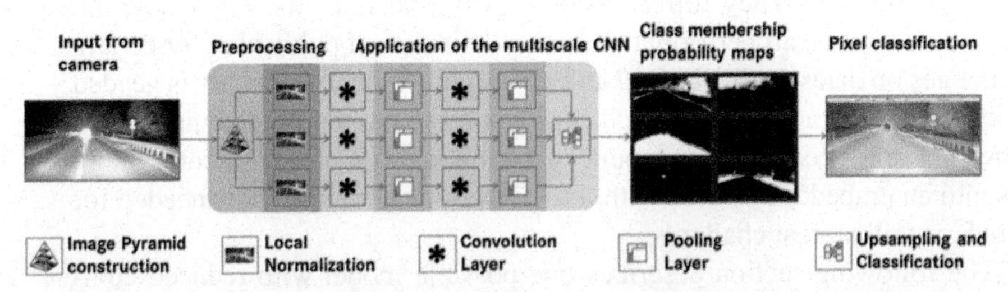

Figure 6.4 The complete processing chain from input image to a scene labeled image is displayed. After building an image pyramid of 3 layers and the local normalization every scale is fed to its own processing chain. This produces 6 class membership probability maps. They can be interpreted and augmented as seen in the output image.

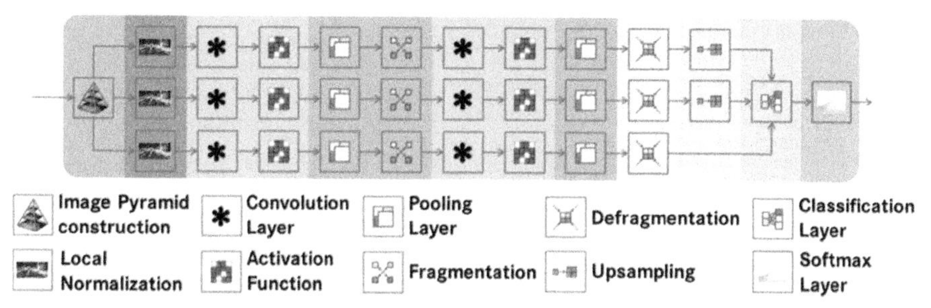

Figure 6.5 The image pyramid construction layer produces 3 scales that are locally normalized in 15×15 windows. Every scale is propagated independently. There are in total 2 convolution layers with $16 \times 7 \times 7$ filter kernels using the ReLU activation function. After activation a 2×2 max-pooling is performed followed by a fragmentation in the first pooling layer. A second fragmentation is not necessary since the second pooling layer is followed by a defragmentation. The small scaled feature maps are sampled up and fed to a classification layer, being a $6 \times 1 \times 1$ convolution layer. Finally, a pixel wise softmax is applied.

It is trained 10.000 epochs with 40960 balanced training examples (patches) per epoch. The learning rate was determined following several short runs of 100 epochs with different learning rates. The best progressing learning rate was then chosen. During training, the learning rate was linearly reduced after 5000 epochs by a factor of 0.995 per epoch. Figure 6.6 shows the training progress (2-2-16 topology) in relation to the objective function on the learn set. Two other topologies were also trained in the same way. One introduced a third convolution layer including the ReLU activation function after the second pooling (3-2-16 topology). The third topology is similar to the 3-2-16 topology, but uses 32 filters per convolution (3-2-32 topology). Figure 6.6 shows that the topology with the least trainable parameters (2-2-16 topology) performed worst during training. The introduction of another convolution layer (3-2-16 topology) resulted in a better learn curve. However, doubling the amount of filters (3-2-32 topology) increased the learn performance yet again.

Since the classifier of topology 3-2-32 appears to have the best performance, it is evaluated on the evaluation set of images containing 200 images that have not been part of the learn set, called the *eval set*. Evaluation in multiclass problems is done by analyzing the *confusion matrix*. The confusion matrix for topology 3-2-32 is displayed in Table 6.1. It shows the class predictions in relation to the actual class. The diagonal entries form the *true positives* (pixels that were classified correctly, TP) for each class, while the remaining entries of a line or column display the individual *false negatives*

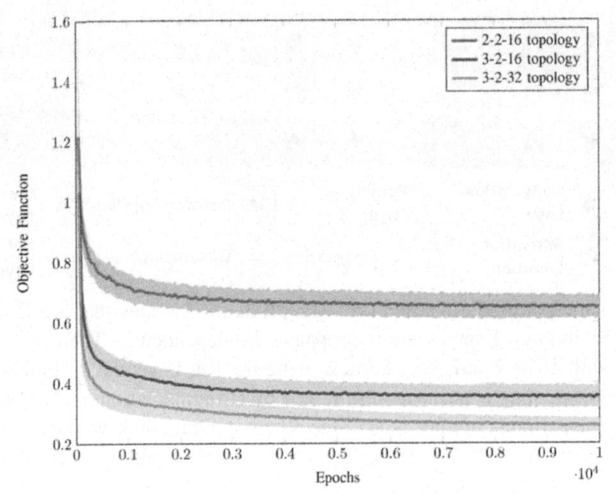

Figure 6.6 Displayed are the learn curves of three different network topologies. Each topology was trained three times and the learn curves were averaged. The averaged learn curves are displayed as solid lines while the standard deviation for 50 epochs is displayed as the area around the lines.

Table 6.1 The confusion matrix of topology 3-2-32 and the respective FNR, FPR and IU for each class. The classes are background (Bg), road (Rd), vehicle (Veh), sky, vulnerable road users (VRU) and infrastructure (Inf). Each cell shows the percentage (from all pixels in the dataset) of actual class (row) predicted as class (column)

Act \ Pred	Bg	Rd	Veh	Sky	VRU	Inf
Bg	24.9349[a]	1.9409	1.1226	2.1282	0.3359[b]	5.8754
Rd	1.5685	29.4059[a]	1.0226	0.0034	0.1269	0.3226
Veh	0.1042	0.0829	3.6523[a]	0.0051	0.1156	0.7749
Sky	1.7298	0.0080	0.1744	7.1476[a]	0.0083	0.9632
VRU	0.0058	0.0032	0.0740[b]	0.0001	0.0733[a]	0.0777[b]
Inf	1.6244	0.0459	1.0077	0.3351	0.3538[b]	12.8450[a]
FNR	31.38	**9.38**[c]	22.87[d]	28.75	68.68[b]	20.77
FPR	16.79	**6.61**[c]	48.22[d]	25.70	92.77[b]	38.42
IU	60.27	**85.16**[c]	44.89[d]	57.17	6.24[b]	53.02

(pixels not classified as the desired class, FN) and *false positives* (pixels falsely classified as the desired class, FP). Therefore, the sum over one row of the table gives the percentage of the respective class in the whole training set.

The quality measures of binary classification problems can therefore be applied for each class individually in a "one versus all" fashion. Classic measures contain the *False Negative Rate* (FNR), the *False Positive Rate* (FPR) and the *Intersection over Union* (IU). Those are defined as follows:

$$FNR = \frac{FN}{N}, \quad FPR = \frac{FP}{N}, \quad IU = \frac{TP}{TP \cup FP \cup FN}$$

N denotes the number of all pixels evaluated. FNR and FPR are 0, if the classification is correct and get bigger, if more pixels are classified incorrectly. The IU has a value of 1 in case of a perfect classification and the value gets smaller, if more pixels are classified incorrectly.

Table 6.1 shows the percentage of pixels classified as one of the 6 classes. The last 3 rows display the class-wise FNR, FPR and IU. The confusion matrix shows several interesting features:

a. The diagonal entries show the true positives, the correctly classified pixels. Since the total amount of pixels in the evaluation dataset for each class varies, the maximum possible number for each entry varies as well.
b. For the class *vulnerable road users* (VRU) the classifier performs badly. There are more pixels classified as *vehicles (Veh)* or *infrastructure (Inf)* than VRUs, resulting in a bad FNR. Even worse is the FPR, since the amount of *background (Bg)* or *infrastructure (Inf)* pixels classified as VRU is far greater than the amount of correctly classified pixels. This results in a bad IU.
c. The best performing class is the class *road (Rd)*. It has comparatively few false positives and negatives, which results in a good FNR, FPR and IU.
d. The class *vehicle (Veh)* shows an arbitrary performance. Though the FNR is quite good and better than the class *background (Bg)*, its FPR is second to last. So the IU is greatly affected.

After analyzing each class by itself the question arises of how good this classifier is compared to classifiers, which contain other well and bad performing classes. A common measure to describe the overall performance of a classifier is the *accuracy* (ACC). It is the ratio of correctly classified pixels to all pixels. Let N be the amount of classes and $C_{i,j}$ be the amount of pixels from class i classified as class j. In a multiclass setup, the accuracy can then be defined as:

$$\text{ACC} = \frac{\sum_{k=1}^{N} C_{k,k}}{\sum_{i,j=1}^{N} C_{i,j}}$$

This measure captures in a straight forward way the correctness of a classifier. The value is in the range [0, 1], where a perfect classifier reaches 1. If one or more classes are under-represented in the evaluation dataset, the expressiveness of this measure suffers, since it does not normalize the amount of samples per class. Other ways to increase the sensitivity to underperforming classes is to average the FNR, FPR or IU over the classes. The *Matthews Correlation Coefficient* (MCC) was designed for binary classifications and computes a correlation between the actual and predicted classifications. It was extended to incorporate more than two classes and is defined by [35] as follows:

$$\mathrm{MCC} = \frac{\sum_{k,l,m=1}^{N} C_{k,k}C_{m,l} - C_{l,k}C_{k,m}}{\sqrt{\sum_{k=1}^{N}\left[\left(\sum_{l=1}^{N} C_{l,k}\right)\left(\sum_{\substack{f,g=1\\ f\neq k}}^{N} C_{g,f}\right)\right]\sum_{k=1}^{N}\left[\left(\sum_{l=1}^{N} C_{k,l}\right)\left(\sum_{\substack{f,g=1\\ f\neq k}}^{N} C_{f,g}\right)\right]}}$$

The Matthews Correlation Coefficient is in the range $[-1, 1]$. An MCC of 1 is a perfect classifier, while -1 is the total contradiction. An MCC of 0 is a random classifier. Table 6.2 shows the ACC, mean IU, MCC and mean FNR for the classifiers trained in Figure 6.6. It can be seen that topology 3-2-32 outperforms the topologies in all defined measures.

6.5 Hardware Platforms for Scene Labeling

Embedded hardware platforms for Advanced Driver Assistance Systems face several challenges. They have to provide a huge amount of processing power to keep up with the rising complexity of applications and the increasing amount of data they have to process. However, the platforms should have low power consumption.

At one end of the spectrum of hardware architectures, *General Purpose Processors* (GPPs) usually do not fulfill all the requirements and restrictions of embedded systems in advanced driver assistance systems. They offer a high degree of flexibility due to the arbitrary programmability, but they cannot usually comply with the high demand on processing power while holding the restrictions in power consumption.

Table 6.2 Displayed are the measures Accuracy (ACC), mean Intersection over Union (mIU), Matthews Correlation Coefficient (MCC) and mean False Negative Rate (mFNR) for 3 topologies

Topology	ACC	mIU	MCC	mFNR
2-2-16	0.60	0.35	0.50	0.44
3-2-16	0.69	0.42	0.60	0.37
3-2-32	**0.78**	**0.51**	**0.71**	**0.30**

At the other end of the spectrum, *Application Specific Integrated Circuits* (ASICs) provide a high degree of processing power and excellent power efficiency. However, they are not flexible as they are fixed after manufacturing and cannot be programmed.

There is a wide range of hardware platforms in between these two extremes, which provide a trade-off between the different characteristics. For example, *Graphical Processing Units* (GPUs) have been used to accelerate the execution of complex algorithms. They provide a certain degree of flexibility, as they are programmable and they achieve high processing power due to a high degree of parallelism. However, the power consumption of GPUs is fairly high and they are therefore not suitable for use in personal cars.

Adapting processor architectures to a given application is a promising approach for designing hardware platforms. *Application-Specific Instruction-Set Processors* (ASIPs) are based on programmable processor architectures. These are adapted to a specific application or a class of similar applications, e.g., by extending the instruction set, by adding dedicated hardware accelerators for frequently used operations, or by changing architectural parameters in order to bypass bottlenecks.

Scene labeling has been implemented on several platforms including CPUs, GPUs, FPGAs, and ASICs. This section gives an overview of recent implementations of convolutional neural networks on different types of computing platforms. At first, the computational complexity of convolutional neural networks is discussed, by deriving a measure of the total number of operations needed in order to compute the forward propagation of one frame through the network. This also serves as a basis for the comparison of different implementations, which is presented later.

6.5.1 Theoretical Performance Requirements

This section describes the computational complexity of convolutional neural networks in terms of operations needed in the forward propagation of a frame. This number of operations clearly depends on the topology of the network.

The most computational intensive task is the convolution, especially, as many convolution layers contain a huge number of filters. For an input image of size $w \times h$ and a convolution kernel of size $n \times n$, the kernel is applied $(w-(n-1))(h-(n-1))$ times. Each time, n^2 multiplications are performed and the results accumulated. Counting the multiply and accumulate operations as two, this leads to a total count of

$$N_{conv}(w, h, n) = 2(w - (n - 1))(h - (n - 1))n^2$$

operations for a single convolution.

The activation function is applied to each output pixel of the input layer. Therefore, the total number of operations for an input image of size $w \times h$ is given as

$$N_{act}(w, h, c_{act}) = whc_{act},$$

where c_{act} describes the cost of applying the activation function to one pixel. In case of the ReLU (Rectified Linear Unit), the operation determines the maximum of the input value and 0. Therefore, $c_{ReLU} = 1$.

For the pooling layer, the number of operations depends not only on the size $w \times h$ of the input frame, but also on the kernel size $n \times n$ and the stride s. In some cases, the stride equals the kernel size, but in overlapped pooling, a stride of 1 might be used. In general, the number of operations performed in a pooling layer can be described as

$$N_{pool}(w, h, n, s) = c_{pool} \frac{wh + (s - n)((s - n) + w + h)}{s^2},$$

where c_{pool} is the number of operations per pooling window. For a max-pooling, the number of operations is $c_{\mathrm{max}} = n^2 - 1$, for an average-pooling, the number of operations is $c_{\mathrm{avg}} = n^2 + 1$.

For the exemplary convolutional neural network described in subsection 6.4.1, which is named 2-2-16 in Table 6.2, the following remarks give the numbers of operations for the single layers. The image preprocessing, i.e., the construction of the image pyramid and the normalization, is not counted in this section.

In this exemplary case, the input image has 1024×512 pixels. In the preprocessing step, an image pyramid is generated by an iterative process. In each iteration, the image dimensions are halved by subsampling. Afterwards, the three scaled images from the pyramid are padded by replicating the border pixels in order to maintain the correct output size after the convolutions. The resulting image sizes are listed in Table 6.3.

The first convolution layer performs 16 convolutions with a 7×7 kernel and generates 16 output images. The convolution is only performed for pixels where the convolution kernel fits into the input image, so that the resulting image is reduced by 6 pixels in width and height. The convolution layer is followed by an activation layer, which applies the activation function to each

Table 6.3 Input image sizes for three different scales in the exemplary convolutional neural network

Scale	Pyramid Output	Padded
S	512×256	534×278
M	256×128	278×150
L	128×64	150×86

of the 16 output images of the convolutions. The following max-pooling layer uses a 2×2 patch and a stride of 1 (overlapped pooling). It does not change the total number of pixels but separates one image into four sub images of quarter size. The fragmentation of the images does not contribute to the number of operations since it can be hidden in the other layers. The second convolution layer performs 16 convolutions of size 7×7 on each of the 16 fragmented images and then accumulates them to 16 fragmented output images. The following activation function and pooling layers work the same as after the first convolution layer.

This flow of images through two convolution layers with activation functions and two pooling layers is performed independently for the three scales of the input image. The resulting images are scaled to the same size before they are fed into the classification layer.

The classification layer at the end performs one convolution of size 1×1 per output class, of which there are six in the exemplary convolutional neural network.

With these image and filter sizes, the computational complexity of the convolutional neural network can be estimated using the equations above. Table 6.4 gives the operation counts for the three scales by layer type.

The total number of operations performed for one input image is 4.796.792.784. As expected, the convolution layers contribute the biggest share in the number of operations, with a proportion of 99.2 percent. In order to reach a processing rate of 30 frames per second, 144 billion operations have to be performed per second.

Table 6.4 Number of operations for the exemplary convolutional neural network

Scale	Convolution	Activation	Pooling	Classif.	Operations
S	3.590.995.968	4.444.416	13.220.592	12.582.912	**3.621.243.888**
M	922.435.584	1.175.808	3.470.064	3.145.728	**930.227.184**
L	243.253.248	327.936	954.096	786.432	**245.321.712**
Ops.	**4.756.684.800**	**5.948.160**	**17.644.752**	**16.515.072**	**4.796.792.784**

Table 6.5 lists implementations of convolutional neural networks on different platforms and gives the performance in terms of performed operations per second. When available, two numbers are given for each implementation. The *peak* performance gives the theoretical maximum number of operations per second that the platform can perform. The *real* performance gives the number of operations per second for CNNs of different topologies on the platform. Not all implementations listed in the table are used for scene labeling, but perform other image based detection and classification tasks with convolutional neural networks. Therefore, the networks that are used in the applications may differ in size. This is mentioned, because some implementations do not scale up to bigger networks easily. The subsequent sections give more details to the entries in the table.

Table 6.5 Comparison of different implementations of convolutional neural networks on different platforms

			Perf. [GOPs]	
Author	Year	Device	Peak	Real
		CPU Implementations		
Farabet et al. [39]	2011	Intel Core 2 Duo	10	1.1
Dundar et al. [40]	2013	Intel Core i7 4-core	200	90
Jin et al. [41]	2014	Intel Core i5	45	30
Zhang et al. [42]	2015	Intel Xeon	–	12.87
		GPU Implementations		
Farabet et al. [39]	2011	nVidia GTX 480	1350	294
Dundar et al. [40]	2013	nVidia GTX 780	3977	620
Jin et al. [41]	2014	nVidia GTX 690	5622	530
Cavigelli et al. [43]	2015	nVidia GTX 780	3977	1781
		Mobile GPU Implementations		
Farabet et al. [39]	2011	nVidia GT335m	182	54
Dundar et al. [40]	2013	nVidia GTX650m	182	54
Cavigelli et al. [43]	2015	nVidia Tegra K1	326	76
		FPGA Implementations		
Farabet et al. [39]	2011	Virtex 6 VLX240T	160	147
Dundar et al. [40]	2013	Zync ZC706	–	36
Gokhale et al. [44]	2014	Zync ZC706	–	227
Zhang et al. [42]	2015	Virtex 7 485t	–	61.62
		ASIC Implementations		
Pham et al. [45]	2012	neuFlow in IBM 45 nm	320	294
Chen et al. [46]	2015	Accelerator in 65 nm	–	452
Cavigelli et al. [47]	2015	Accelerator in 65 nm	274	203

6.5.2 CPU-based Platforms

As discussed before, running convolutional neural networks for scene labeling or other image processing tasks incorporates a huge amount of computation. For the use in ADAS, CPUs cannot provide the necessary processing power while also complying to the power budget restrictions. Active work is performed in order to speed up the implementations (e.g., [36]). Also, algorithmic research is conducted in order to speed up the convolutions, e.g., [37, 38].

A reference implementation of the exemplary CNN from subsection 6.4.1 was written using C++. It is worth mentioning that the focus in this implementation was not speed or efficiency. Instead, it was intended as a reference for the assembler implementation described later. The implementations of the image processing operations and the different layers of the convolutional neural network make use of templates. This provides the flexibility to use different data types for the pixel values and coefficients. The templates enabled the use of fixed-point data types in order to analyze the compromise of data width and accuracy.

On an Intel Core i5-2400 with 3.1 GHz, the computations for one input image of size 1024×512 with double precision values and coefficients require about 11 seconds, which corresponds to about 436 MOPS. This implementation does not use multiple cores for computation.

6.5.3 GPU-based Platforms

Modern GPUs provide a huge amount of computing power that can be used for general purpose computing (GPGPU). The use of GPUs is most beneficial, if the application provides a high degree of parallelism and regularity. CNNs fall into this category. Therefore, most deep learning frameworks mentioned in the previous section accelerate evaluation and training of networks with GPUs using CUDA, and there are also frameworks specifically developed for GPUs, e.g., cuda-convnet2 [29] and Marvin [48].

A downside of using the powerful GPUs is the amount of power they consume, which makes the use of GPUs in mobile devices infeasible. Nevertheless, GPUs can be used for training the networks, as the training is performed offline. Recently, mobile or embedded GPUs have emerged, aiming to provide low-power high-performance computing platforms.

6.5.4 FPGA-based Platforms

A FPGA, a configurable hardware platform, provides a compromise between the flexibility of a GPU and the efficiency of an ASIC. The high degree of parallelism that is possible in a FPGA, allows for high performance signal processing. As double precision arithmetic is costly for a hardware-based implementation, the C++ implementation of the algorithm was used to analyze the quality of the classification depending on the data width of pixel values and coefficients. For 32-bit data with 22 fractional bits, the computations are exact and no errors appear. If 16-bit data with 11 fractional bits are used, about 1.4 percent of the pixels are classified incorrectly, which was acceptable in this scenario.

The use of a soft core processor that is mapped to the FPGA also provides software programmability of the design. In order to raise the computational performance, the soft core processor can be extended with dedicated hardware modules (application-specific instruction-set processor, ASIP). For example, the instruction-set can be extended by new functional units for complex operations which are placed in the processor's pipeline and perform as quick as the default operations. Additionally, more complex operations taking more execution cycles can be added as external accelerators tightly coupled with the processor's data path.

In the course of the DESERVE project, an ASIP implementation for convolutional neural networks has been developed. It is based on the TUKUTURI processor [49, 50], which was developed for image processing and video coding implementations. It is a Very Long Instruction Word (VLIW) processor with two issue slots and 64 bit wide registers that can be split up into subwords of 8, 16, 32, or 64 bits. These subwords are processed in parallel (microSIMD) by all default functional units. Additional features include conditional execution in order to reduce control overhead, and a DMA controller for memory transfer between external and internal memory.

As derived from the CPU-based reference implementation (see subsection 6.5.2), 16 bit wide data is used for the pixel values and the network's coefficients. Therefore, the SIMD-feature can be used to process four values in parallel, which gives a significant speed-up.

As seen in subsection 6.5.1, the convolution is the most computing intensive task in the whole process. Therefore, the TUKUTURI processor was extended with a co-processor that performs 16 convolutions of four pixels at once.

The internal memory of the TUKUTURI is not capable of holding a whole input image. Therefore, the images are processed in blocks. The DMA module supports block transfers, so that a rectangular subsection of the image can be transferred between internal and external memory. The module holds a queue of memory transfers, which are processed independently from the TUKUTURI processor. This allows the TUKUTURI to program several transfers and process data blocks transferred previously, while the DMA transfers the next blocks in the background.

The first implementation of the exemplary convolutional neural network on the TUKUTURI processor processed one input frame in about 1.2×10^9 cycles. With a clock frequency of 100 MHz, this corresponds to about 0.08 fps. Using the convolution co-processor, the cycle count could be reduced to about 243×10^6 cycles, corresponding to a frame rate of about 0.411 fps. This is a speed-up of factor 5.1. Using the capabilities for background transfers, the total cycle count was reduced to about 101×10^6 cycles per frame, which is an additional speed-up of factor 2.4, leading to about 0.99 fps. According to Table 6.4, we need about 4.8×10^9 operations per frame. Therefore, this implementation reaches about 4.8 GOPs.

6.6 Summary

Convolutional neural networks and methods of deep learning have been used in image processing, segmentation and classification tasks successfully. The huge amount of processing power needed for CNNs for Scene Labeling tasks in advanced driver assistance systems combined with the resource restrictions in embedded systems pose a challenge for hardware architects. FPGAs have been shown as a suitable platform for the implementation of CNNs for Scene Labeling.

References

[1] G. Carneiro and N. Vasconcelos, "Formulating semantic image annotation as a supervised learning problem," in *2005 IEEE Computer Society Conference on Computer Vision and Pattern Recognition (CVPR'05)*, 2005, pp. 163–168.

[2] E. Saber, A. Tekalp, R. Eschbach and K. Knox, "Automatic Image Annotation Using Adaptive Color Classification," *Graphical Models and Image Processing,* 1996.

[3] J. Shotton, J. Winn, C. Rother and A. Criminisi, "TextonBoost for Image Understanding: Multi-Class Object Recognition and Segmentation by Jointly Modeling Texture, Layout, and Context," *International Journal of Computer Vision,* 2009.

[4] M. Pietikäinen, T. Nurmela, T. Mäenpää and M. Turtinen, "View-based recognition of real-world textures," *Journal of Pattern Recognition,* 2004.

[5] X. Ren, L. Bo and D. Fox, "RGB-(D) scene labeling: Features and algorithms," *Computer Vision and Pattern Recognition (CVPR)*, 2012.

[6] P. F. Felzenszwalb and O. Veksler, "Tiered scene labeling with dynamic programming," *Computer Vision and Pattern Recognition (CVPR),* 2010.

[7] S. M. Bhandarkar and H. Zhang, "Image segmentation using evolutionary computation," *IEEE Transactions on Evolutionary Computation,* 1999.

[8] A. Ess, B. Leibe, K. Schindler and L. V. Gool, "A mobile vision system for robust multi-person tracking," *Computer Vision and Pattern Recognition,* 2008.

[9] A. Broggi, P. Cerri, P. Medici, P. P. Porta and G. Ghisio, "Real Time Road Signs Recognition," *2007 IEEE Intelligent Vehicles Symposium,* 2007.

[10] J. C. McCall and M. M. Trivedi, "Video-based lane estimation and tracking for driver assistance: survey, system, and evaluation," *Intelligent Transportation Systems,* 2006.

[11] B. Fulkerson, A. Vedaldi and S. Soatto, "Class segmentation and object localization with superpixel neighborhoods," in *Computer Vision, 2009 IEEE 12th International Conference on*, 2009.

[12] A. Torralba, K. P. Murphy and W. T. Freeman, "Sharing Visual Features for Multiclass and Multiview Object Detection," *Pattern Analysis and Machine Intelligence, IEEE Transactions on,* Vol. 29, No. 5, pp. 854–869, May 2007.

[13] M. Turtinen and M. Pietikäinen, "Contextual Analysis of Textured Scene Images," *British Machine Vision Conference,* 2006.

[14] B. Hariharan, P. Arbelaez, R. Girshick and J. Malik, "Simultaneous Detection and Segmentation," *Computer Vision – ECCV,* 2014.

[15] X. He, R. S. Zemel and M. A. Carreira-Perpinan, "Multiscale conditional random fields for image labeling," in *Computer Vision and Pattern Recognition, 2004. CVPR 2004. Proceedings of the 2004 IEEE Computer Society Conference on*, 2004.

[16] X. Liu, O. Veksler and J. Samarabandu, "Order-Preserving Moves for Graph-Cut-Based Optimization," *Pattern Analysis and Machine Intelligence, IEEE Transactions on,* Vol. 32, No. 7, pp. 1182–1196, July 2010.

[17] W. S. McCulloch and W. Pitts, "A logical calculus of the ideas immanent in nervous activity," *The bulletin of mathematical biophysics,* Vol. 5, No. 4, pp. 115–133, 1943.
[18] F. Rosenblatt, "The perceptron: A probabilistic model for information storage and organization in the brain," *Psychological Review,* Vol. 65, No. 6, 1958.
[19] C. von der Malsburg, "Self-organization of orientation sensitive cells in the striate cortex," *Kybernetik,* Vol. 14, No. 2, pp. 85–100, 1973.
[20] J. J. Hopfield, "Neural networks and physical systems with emergent collective computational abilities," *Proceedings of the national academy of sciences,* Vol. 79, No. 8, pp. 2554–2558, 1982.
[21] X. Glorot, A. Bordes and Y. Bengio, "Deep sparse rectifier neural networks," *Proceedings of the 14th International Conference on Artificial Intelligence and Statistics,* 2011.
[22] D. E. Rumelhart, G. E. Hinton and R. J. Williams, "Lerning representations by back-propagating errors," in *Nature*, Vol. 323, Nature Publishing Group, 1986, pp. 533–536.
[23] K. Fukushima and S. Miyake, "Neocognitron: A new algorithm for pattern recognition tolerant of deformations and shifts in position," *Pattern Recognition,* Vol. 15, No. 6, pp. 455–469, 1982.
[24] Y. Lecun, L. Bottou, Y. Bengio and P. Haffner, "Gradient-based learning applied to document recognition," *Proceedings of the IEEE,* Vol. 86, No. 11, pp. 2278–2324, Nov 1998.
[25] A. Giusti, D. Ciresan, J. Masci, L. Gambardella and J. Schmidhuber, "Fast image scanning with deep max-pooling convolutional neural networks," in *Image Processing (ICIP), 2013 20th IEEE International Conference on*, 2013.
[26] Y. Jia, E. Shelhamer, J. Donahue, S. Karayev, J. Long, R. Girshick, S. Guadarrama and T. Darrell, "Caffe: Convolutional Architecture for Fast Feature Embedding," in *Proceedings of the 22nd ACM International Conference on Multimedia*, New York, NY, USA, 2014.
[27] R. Collobert, K. Kavukcuoglu and C. Farabet, "Torch7: A Matlab-like Environment for Machine Learning," in *BigLearn, NIPS Workshop*, 2011.
[28] J. Bergstra, O. Breuleux, F. Bastien, P. Lamblin, R. Pascanu, G. Desjardins, J. Turian, D. Warde-Farley and Y. Bengio, "Theano: A CPU and GPU Math Compiler in Python," in *9th Pytthon in Science Conference (SCIPY 2010), Proceedings of the*, 2010.

[29] A. Krizhevsky, "cuda-convnet2," 2014. [Online]. Available: https://code.google.com/archive/p/cuda-convnet2/. [Accessed März 2016].

[30] A. Krizhevsky, I. Sutskever and G. E. Hinton, "ImageNet Classification with Deep Convolutional Neural Networks," in *Advances in Neural Information Processing Systems 25*, F. Pereira, C. Burges, L. Bottou and K. Weinberger, Eds., Curran Associates, Inc., 2012, pp. 1097–1105.

[31] C. Szegedy, W. Liu, Y. Jia, P. Sermanet, S. Reed, D. Anguelov, D. Erhan, V. Vanhoucke and A. Rabinovich, "Going Deeper with Convolutions," in *CVPR 2015*, 2015.

[32] K. Simonyan and A. Zisserman, "Very Deep Convolutional Networks for Large-Scale Image Recognition," *CoRR,* vol. abs/1409.1556, 2014.

[33] O. Russakovsky, J. Deng, H. Su, J. Krause, S. Satheesh, S. Ma, Z. Huang, A. Karpathy, A. Khosla, M. Bernstein, A. Berg and L. Fei-Fei, "ImageNet Large Scale Visual Recognition Challenge," *International Journal of Computer Vision,* Vol. 115, No. 3, pp. 211–252, 2015.

[34] C. Farabet, C. Couprie, L. Najman and Y. LeCun, "Learning Hierarchical Features for Scene Labeling," *IEEE Transactions on Pattern Analysis and Machine Intelligence,* Vol. 35, No. 8, pp. 1915–1929, 2013.

[35] G. Jurman, S. Riccadonna and C. Furlanello, "A Comparison of MCC and CEN Error Measures in Multi-Class Prediction," *PLoS ONE,* Vol. 7, No. 8, p. e41882, 08 2012.

[36] F. Bastien, P. Lamblin, R. Pascanu, J. Bergstra, I. J. Goodfellow, A. Bergeron, N. Bouchard, D. Warde-Farley and Y. Bengio, "Theano: new features and speed improvements," *CoRR,* vol. abs/1211.5590, 2012.

[37] V. Lebedev, Y. Ganin, M. Rakhuba, I. V. Oseledets and V. S. Lempitsky, "Speeding-up Convolutional Neural Networks Using Fine-tuned CP-Decomposition," *CoRR,* vol. abs/1412.6553, 2014.

[38] J. Cong and B. Xiao, "Minimizing Computation in Convolutional Neural Networks," in *Artificial Neural Networks and Machine Learning – ICANN 2014: 24th International Conference on Artificial Neural Networks, Hamburg, Germany, September 15-19, 2014. Proceedings*, S. Wermter, C. Weber, W. Duch, T. Honkela, P. Koprinkova-Hristova, S. Magg, G. Palm and A. E. P. Villa, Eds., Springer International Publishing, 2014, pp. 281–290.

[39] C. Farabet, B. Martini, B. Corda, P. Akselrod, E. Culurciello and Y. LeCun, "NeuFlow: A runtime reconfigurable dataflow processor for vision," in *Computer Vision and Pattern Recognition Workshops (CVPRW), 2011 IEEE Computer Society Conference on*, 2011.

[40] A. Dundar, J. Jin, V. Gokhale, B. Krishnamurthy, A. Canziani, B. Martini and E. Culurciello, "Accelerating deep neural networks on mobile processor with embedded programmable logic," in *Neural information processing systems conference (NIPS)*, 2013.

[41] J. Jin, V. Gokhale, A. Dundar, B. Krishnamurthy, B. Martini and E. Culurciello, "An efficient implementation of deep convolutional neural networks on a mobile coprocessor," in *Circuits and Systems (MWSCAS), 2014 IEEE 57th International Midwest Symposium on*, 2014.

[42] C. Zhang, P. Li, G. Sun, Y. Guan, B. Xiao and J. Cong, "Optimizing FPGA-based Accelerator Design for Deep Convolutional Neural Networks," in *Proceedings of the 2015 ACM/SIGDA International Symposium on Field-Programmable Gate Arrays*, New York, NY, USA, 2015.

[43] L. Cavigelli, M. Magno and L. Benini, "Accelerating Real-time Embedded Scene Labeling with Convolutional Networks," in *Proceedings of the 52nd Annual Design Automation Conference*, New York, NY, USA, 2015.

[44] V. Gokhale, J. Jin, A. Dundar, B. Martini and E. Culurciello, "A 240 G-ops/s Mobile Coprocessor for Deep Neural Networks," in *Computer Vision and Pattern Recognition Workshops (CVPRW), 2014 IEEE Conference on*, 2014.

[45] P.-H. Pham, D. Jelaca, C. Farabet, B. Martini, Y. LeCun and E. Culurciello, "NeuFlow: Dataflow vision processing system-on-a-chip," in *Circuits and Systems (MWSCAS), 2012 IEEE 55th International Midwest Symposium on*, 2012.

[46] T. Chen, Z. Du, N. Sun, J. Wang, C. Wu, Y. Chen and O. Temam, "A High-Throughput Neural Network Accelerator," *Micro, IEEE,* Vol. 35, No. 3, pp. 24–32, May 2015.

[47] L. Cavigelli, D. Gschwend, C. Mayer, S. Willi, B. Muheim and L. Benini, "Origami: A Convolutional Network Accelerator," in *Proceedings of the 25th Edition on Great Lakes Symposium on VLSI*, New York, NY, USA, 2015.

[48] "Marvin: A minimalist GPU-only N-dimensional ConvNet framework," [Online]. Available: http://marvin.is. [Accessed 2015].

[49] G. Payá-Vayá, R. Burg and H. Blume, "Dynamic Data-Path Self-Reconfiguration of a VLIW-SIMD Soft-Processor Architecture," *Workshop on Self-Awareness in Reconfigurable Computing Systems (SRCS) in conjunction with the 2012 International Conference on Field Programmable Logic and Applications (FPL 2012)*, 2012.

[50] S. Nolting, G. Payá-Vayá and H. Blume, "Optimizing VLIW-SIMD Processor Architectures for FPGA Implementation," *Proceedings of the ICT.OPEN 2011 Conference* (Veldhoven, Netherlands), 2011.
[51] A. Ess, T. Mueller, H. Grabner and L. v. Gool, "Segmentation-based urban traffic scene understanding," in *Proceedings of the British Machine Vision Conference*, 2009.
[52] Y. LeCun, K. Kavukcuoglu and C. Farabet, "Convolutional networks and applications in vision," in *Circuits and Systems (ISCAS), Proceedings of 2010 IEEE International Symposium on*, 2010.

7

Real-Time Data Preprocessing for High-Resolution MIMO Radar Sensors

Frank Meinl[1], Eugen Schubert[1], Martin Kunert[1] and Holger Blume[2]

[1]Advanced Engineering Sensor Systems, Robert Bosch GmbH, Leonberg, Germany
[2]Institute of Microelectronic Systems, Leibniz Universität Hannover, Hannover, Germany

7.1 Introduction

The progress in resolution of automotive radar sensors involves a considerable increase in data-rate and computational throughput. Dedicated processing architectures have to be investigated in order to manage the tremendous amount of data. Even for early prototype development platforms, the performance of existing PC-based frameworks and tools is no longer sufficient to cope with the data processing of many parallel radar receiver channels at very high sampling rates.

This chapter presents a FPGA-based signal processing architecture capable of handling 16 parallel MIMO radar receiving channels with a sampling frequency of 250 MHz each. Raw data is transferred from the AD-Converters to the FPGA where subsequent processing steps are performed, involving FIR-filtering and decimation, two-dimensional FFT transform, local noise level estimation and subsequent target detection. An external DRAM is used for storing multiple radar measurements which are finally evaluated altogether (so-called chirp-sequence modulation).

Data post-processing is outsourced onto a PC running with ADTF, an automotive framework for graph-based real-time data processing. The combination of a fast, FPGA-based preprocessing unit with a more flexible, PC-based development platform maximizes processing performance and minimizes development time. The less mature angular MIMO processing algorithms can thus be evaluated with the help of C-based algorithms running in ADTF, while the simple, but calculation intensive FFT processing is implemented entirely as a hardware accelerator in a Virtex-7 FPGA device from Xilinx.

7.2 Signal Processing for Automotive Radar Sensors

After AD-conversion, the raw radar signals enter the processing unit, consecutively passing through all necessary signal processing steps. Different levels of data abstraction and representation can be identified, which range from low level time signals up to complex environmental models.

In this chapter, only the extraction of discrete scattering centers will be considered. The result is a list of reflections, each having multiple features, like for instance Cartesian coordinates, radar-cross-section (RCS), relative velocity or signal-to-noise ratio. Further processing of these reflections would incorporate clustering, classification and environment modeling.

An intermediate state is the extraction of relevant targets from the two-dimensional frequency spectrum (cf. Subsection 7.2.2). At this point, the range and velocity of the targets have already been determined, while the angular information is not evaluated yet. Nevertheless, the data rates are already reduced by a significant amount, so that at this stage the data transfer interface between FPGA and PC-based signal processing can be established.

7.2.1 FMCW Radar System Architecture

The usage of frequency-modulated continuous-wave (FMCW) radar sensors can be advantageous in short range applications, especially due to their high range resolution capability and much lower peak power requirements. In contrast to a pulsed radar system, the transmitter and receiver operate at the same time, which imposes some constraints on the transmitted signals. In order to measure the time-of-flight, i.e. the range towards an object, some kind of time-varying information needs to be added to the transmitted waveforms. The signal has to be modulated in an unambiguous, non-repetitive fashion. A constant sine wave, for instance, can't be used for range estimation, due to its ambiguity after the phase has increased by one cycle or 2π, respectively.

One widely used modulation scheme consists of linear modulated frequency chirps (cf. Figure 7.1). Two important parameters are the used bandwidth F and the modulation time T which determine the slope $\frac{F}{T}$ of

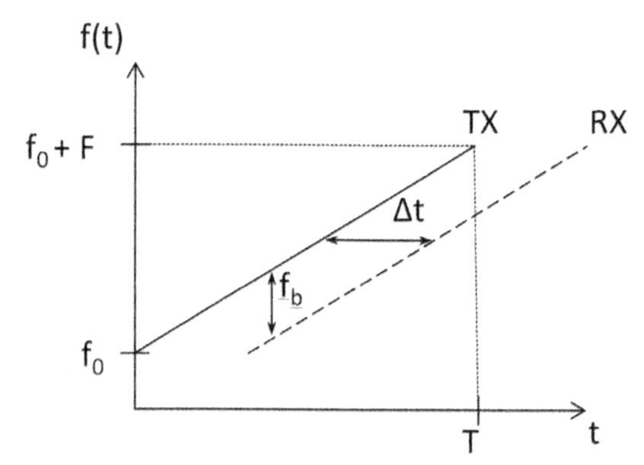

Figure 7.1 FMCW ramp waveform shown as frequency over time f(t). The solid line represents the transmitted signal (TX) while the dashed line is the received signal (RX).

the frequency ramp. Besides, other kinds of modulation schemes exist, e.g. frequency shift keying, various phase modulation or pseudo-noise coding principles.

In the case of a linear frequency modulation, the time-of-flight Δt can be directly translated into a frequency difference (so called beat frequency f_b). With the help of a mixer device in the receiver, this frequency difference can be measured efficiently and estimated by subsequent signal processing blocks. Finally, the target range r can be obtained from the estimated beat frequency value. However, as moving targets engender an additional frequency shift f_d (Doppler frequency), the measured frequency will consist of a superposition of a range and a velocity dependent component.

$$f_b = f_r + f_d = \frac{2r}{c}\frac{F}{T} - \frac{2v_r}{\lambda}$$

With the help of advanced modulation waveforms, the occurrence of range-Doppler ambiguities can be significantly reduced, while being able to estimate both frequency components individually at the same time [1]. This can be achieved by using multiple, aligned FMCW chirps. Furthermore, these ramp signals should have a very steep slope, so that the range dependent frequency part f_r dominates in the beat frequency f_b. For a sufficient small target velocity, the Doppler frequency f_d is likewise small enough so that the range estimation can be carried out directly from f_b by simply neglecting the minor f_d contribution. However, the Doppler information is not completely lost and can be regained from the inherent phase measurement which is present in the consecutive frequency ramps. For this purpose, it is necessary that the ramp sequence is strictly aligned and that the data sampling occurs always at the same time instant w.r.t. the chirp modulation. The underlying processing technique is shown in Figure 7.2 and relies on a two-dimensional spectrum analysis. The big advantage is the unambiguous determination of both the range and velocity frequency component of each target.

For the angle estimation, two different measurement principles can be used. One possibility is a steerable antenna, which has a high directivity. Only targets which reside inside the antenna beam will contribute to the received signal in a significant manner. The detection space has to be scanned individually, i.e. each possible direction of arrival (DOA) will be measured separately. An alternative to a mechanical steered antenna is the use of an antenna array, where each antenna element is fed by a time delayed version of the transmit signal. The phase shift of the antenna feeds can be changed electronically. Depending on the phase relationships of the antenna elements, the directivity can be swiveled, which is also referred to as electronic beam steering or phased array.

The second class of angle estimation relies on a phase measurement of the received signals. Within a static antenna array, the measured phase differences will depend on the DOA of the target reflections. This property is exploited by many different algorithms in the field of array processing [2]. A major

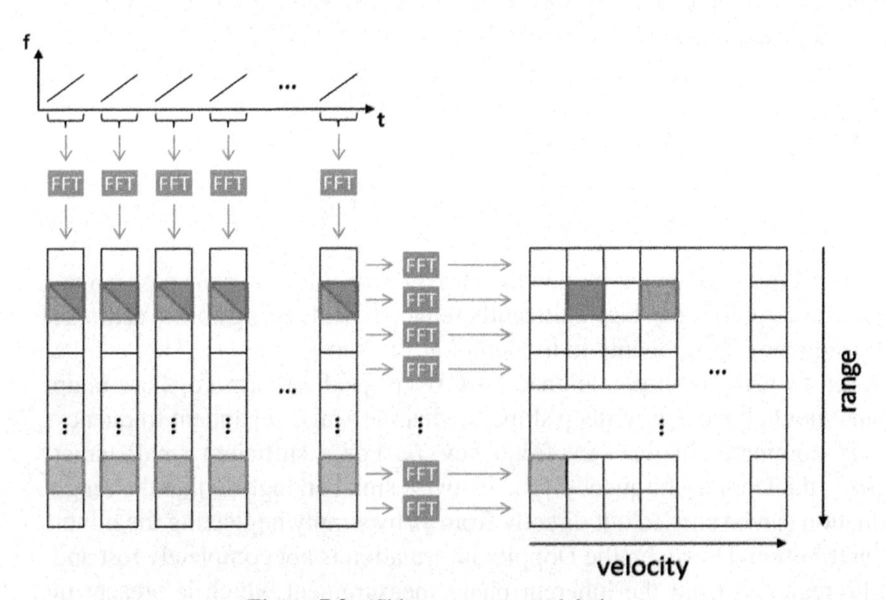

Figure 7.2 Chirp-sequence modulation.

advantage of a fixed antenna array is the simultaneous measurement over a wide opening angle. The region of interest does not have to be scanned and data can be collected in a single, instantaneous snapshot. In general, the achievable angular resolution and separability depends on the number of channels as well as on the aperture size of the array.

In the case of a receiving array, each channel will require a dedicated frequency mixer, amplifier and AD-converter, which increases the total cost of the system. Hence, the usage of advanced algorithms can be considered in order to increase resolution without additional receiving channels [3]. These algorithms are often said to achieve a superresolution because they perform better than a conventional Bartlett beamscan algorithm (cf. [2], pp. 1142). Another possibility is the usage of multiple transmitting channels (multiple input – multiple output – MIMO). A MIMO system has a better efficiency because the number of virtual channels is larger than the real number of channels, thus resulting in lower hardware effort.

In Figure 7.3, a linear MIMO antenna array is shown with two transmitter antennas, which are depicted as circles on the left. The physical receiving array (blue) is extended by several virtual antenna positions. The underlying signal processing remains the same as in the single transmitter case, however the full virtual array can be used resulting in an increased accuracy and object separation capability. In order to separate the signals originating from different transmitting antennas at the receiver side, some kind of orthogonality has to be introduced. A straight forward approach is to use a time-division multiplexing (TDM) approach, i.e. only one transmitter operates at the same time. Other possible techniques comprise frequency-division multiplexing (FDM) or code-division multiple access (CDMA).

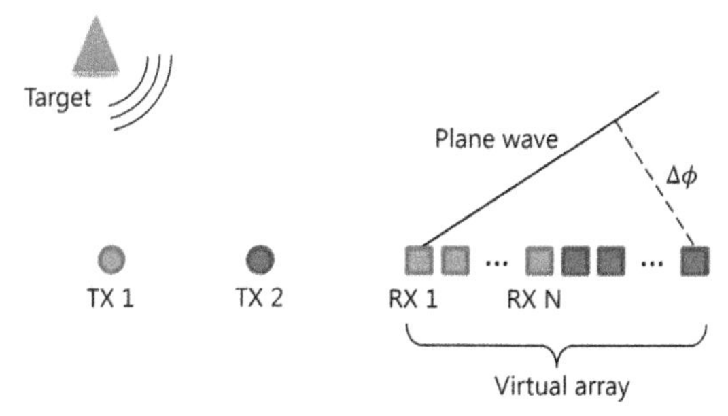

Figure 7.3 Possible MIMO antenna array design: The physical receiver array (blue) is extended by several virtual antennas (red squares) due to the second transmitter TX 2.

7.2.2 Two-Dimensional Spectrum Analysis for Range and Velocity Estimation

Multi-target scenarios are usually encountered in automotive radar applications. Especially static targets are often present in the field of view arising from roadside structures, e.g. guardrails and reflector posts. Furthermore, with increased resolution, multiple scattering centers are visible from single objects, e.g. the shape of car bodies is seen as a large cloud consisting of many reflections [4].

In order to resolve and separate proximate targets, a good range resolution and thus frequency resolution is required. One widely used technique providing a fast and robust frequency estimation is the fast Fourier transform (FFT). For a further increase in range resolution, advanced frequency estimation algorithms like autoregressive (AR) models or multiple signal classification (MUSIC) can be employed [5, 6]. Beside the higher computational requirements, they suffer from the fact that the number of detections needs to be known prior to the estimation. For this reason, the presented system relies on the more convenient FFT-based spectrum analysis.

The Doppler frequency estimation is carried out by a second FFT. Instead of the raw time signals, the frequency bins of the first FFT are used as input signal. In other words, the second FFT measures the ramp-to-ramp phase offset for each target. This offset depends solely on the Doppler shift of the target, because the radar system ensures a coherent sampling of the transmitted frequency chirps. Only if the target is moving relatively to the sensor, the measured phase value will vary between the consecutive chirp ramps.

As depicted in Figure 7.2, targets with different ranges and different velocities are separated after this step. In contrast to many other FMCW modulation forms, a matching step to find corresponding ranges and velocities is no longer required, because the values are directly obtained from the two-dimensional indices. Furthermore, the computational effort stays constant and is thus independent from the number of prevailing targets. This property plays a key role in scenarios with many scattering points as often encountered with high resolution automotive radar sensors.

Another benefit of the two-dimensional spectral processing is the higher sensitivity. Particularly small targets with a low radar cross-section (RCS) can be masked by the noise floor of the first FFT. These targets become visible only by the help of the additional processing gain of the second FFT. Thus, each output bin of the first FFT shall be taken into account and the full 2D matrix should be evaluated before any target detection takes place.

7.2.3 Thresholding and Target Detection

A crucial point in the signal processing chain is the separation of different target reflections in the two-dimensional power spectrum. With the help of this step, data of relevant objects will be isolated from the random noise components. This leads to a significant reduction of data rate and thus lowers the computational performance requirements for the downstream signal processing steps.

The target detection is carried out with the help of an adaptive threshold, reducing the effects of local noise and clutter components. With the means of a constant false alarm rate (CFAR) processing, the probability of false alarm remains constant, irrespective of varying operational and environmental conditions.

Different types of CFAR processors can be used for noise level estimation. Two variants are presented in this section, the cell-averaging (CA-CFAR) and the ordered-statistic (OS-CFAR), two of the most extensively used variants.

Cell-Averaging CFAR (CA-CFAR)

The basic task of a CFAR detector is to provide an adaptive threshold, which is then used for the subsequent detection step, i.e. the decision if a specific cell contains a present target or just irrelevant noise components. In contrast to a fixed threshold, an estimate of the local background noise level is used as threshold, which has to be obtained automatically and separately for each cell under test (CUT). Many different methods exist to provide such an estimate, each leading to different classes and variants of CFAR detectors.

A simple yet powerful approach is the mean value of a number of window cells in proximity to the CUT (see Figure 7.4). This variant is known as cell

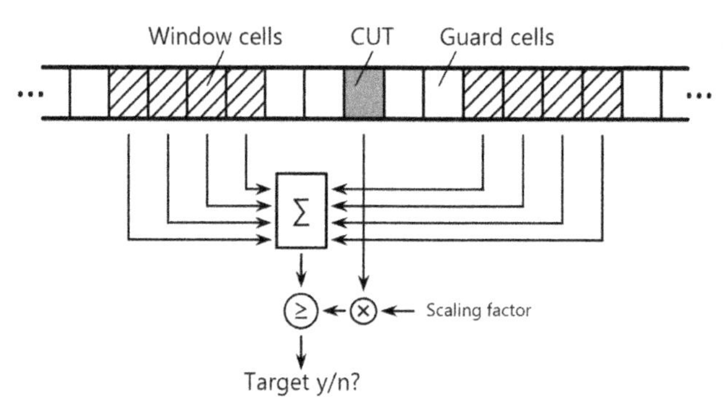

Figure 7.4 CA-CFAR sliding window implementation.

averaging CFAR, or CA-CFAR. The assumption made in this case is that all window cells contain only noise components and thus the mean value is a good estimate of the noise variance. In the case of white Gaussian noise, the value is corresponding to the maximum likelihood estimator. However, for many radar systems the assumption of normal distributed noise turns out to be inaccurate [7].

When designing a CFAR detector, an important parameter is the window size around the CUT. On the one hand, a larger window size reduces the statistical estimation error; on the other hand, local differences in the noise level can be blurred by a large window. A tradeoff has to be made between the deviation from the requested false alarm rate due to the estimation error and the local sensitivity of the adaptive threshold which results from smoothing. Furthermore, the computational effort becomes more relevant with increasing window sizes.

Ordered-Statistic CFAR (OS-CFAR)

In the case of white Gaussian noise, the CA-CFAR performs very well in single target scenarios. However, in a multi-target environment, the estimated noise level will deviate due to interfering targets inside the window cells. Robust statistics can be used in order to suppress outliers arising from other targets inside the window. A commonly used variant is the ordered-statistic (OS-CFAR) which relies on a sortation of the values inside the window, similar to a median filter.

The algorithm performs the following steps for each cell under test (CUT):

- Sort all cells inside the window by their absolute square value
- Take out the k-th value of the sorted list. This value serves as an estimate for the local noise level
- Apply a scaling factor to the noise estimate
- Compare the scaled estimated noise value against the CUT
- Decide whether the CUT is a valid target

Especially in the field of high-resolution radar, big window sizes are required, because large and widespread targets will easily occupy multiple window cells. The complete sortation of the whole window is not a very efficient solution. Only a single value of the sorted list is of interest, while all other values are discarded. Furthermore, when evaluating neighboring CUTs, the previously sorted list can be used as starting point.

Several optimizations of the algorithm aim at these specific sortation characteristics. For instance, a "k-th maximum search" can be performed which finds the greatest value and removes it from the set. This step is repeated until the k-th value has been found [8]. Another efficient realization uses a sliding window approach which keeps a sorted list in memory [9]. Now, when moving the window one step further, the insertion of a single value requires at most N comparisons.

Besides, if one is only interested in the decision result, the complete sortation of the list can be bypassed and the detection step can be performed in a "rank-only" manner [10]. Therefore, the inverse threshold is applied to the CUT and the result is compared to each cell inside the window. The binary comparison results, i.e. 1 if the value is bigger – 0 if not, can be summed up to get a rank. Only if the rank is greater than k, the CUT is considered as valid detection. This approach is depicted in Figure 7.5.

In contrast to a complete sortation, this algorithm depends only on N comparisons. The complexity is thus linear for growing window sizes. The target decision result is exactly the same, i.e. there is no performance loss. The only disadvantage is the lack of the k-th value, which is unknown in the rank-only case. This value can serve as an estimate for the local noise level and can be required by subsequent signal processing blocks. A supplementary estimation of this value can be considered, e.g. the mean value of all cells which have been classified as noise.

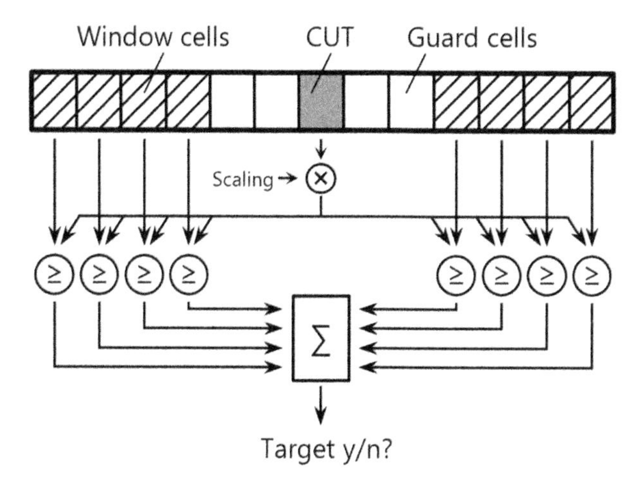

Figure 7.5 Rank-only OS-CFAR implementation.

Non-Coherent Integration (NCI)

Even though the detection takes place before the angular processing, the data of multiple receiving channels can be used to further improve detection performance. An integration of all channels prior to the detection step turns out to be beneficial, assuming that the noise components are independent and identically distributed (i.i.d.). However, the phase relationship of the signals between adjacent channels is not known prior to the angle estimation and can take any value. When summing up the complex values of each channel, the signals can interfere either constructively or destructively. In order to avoid a cancellation of the signal power, the integration takes place in the power spectra, which is also known as non-coherent integration (NCI).

In the following, the noise components are modeled as additive-white Gaussian noise which means that a zero-mean normal distributed signal n[t] is added to the received signal s[t].

It can be shown, that both the real and imaginary parts of the noise components follow a zero-mean normal distribution after transformation into the frequency space [11]. The variance of N[k] depends on the input variance as well as on the length of the input signal, i.e. the length of the FFT. When taking longer signal sequences, the signal-to-noise ratio can be improved (so-called processing gain).

$$\hat{s}[t] = s[t] + n[t]$$

$$\circ\!\!-\!\!\bullet$$

$$\hat{S}[k] = S[k] + N[k]$$

The power spectrum can be calculated by summing up the squared values of real and imaginary part. As a sum of two squared, i.i.d. Gaussian variables, it results a chi-squared distribution $\chi^2(n)$ with $n = 2$ degrees of freedom for the squared magnitude $|N[k]|^2$:

$$|N[k]|^2 = N_{Re}[k]^2 + N_{Im}[k]^2$$
$$|N[k]|^2 \sim \chi^2(2)$$

When summing up multiple receiving channels, i.e. multiple i.i.d. random variables, the result will again be chi-squared distributed but with a higher degree of freedom.

$$N_{NCI}[k] = \sum_{i=1}^{m} |N_i[k]|^2 \sim \chi^2(2m)$$

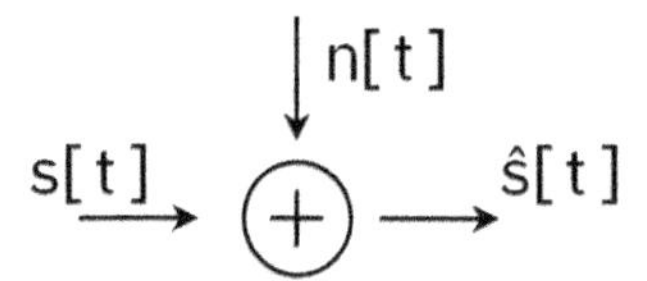

Figure 7.6 Additive white Gaussian noise model.

In contrast to the FFT, the mean value of the noise power scales linearly with the number of channels in the same way the signal power does. Therefore, the signal to noise ratio is not improved. However, the variance is decreasing which has an effect on the possibility of false alarm. An example measurement is depicted in Figure 7.7, comparing the noise distribution of one channel and the distribution after the integration of 32 channels. It can be observed that for the same threshold level, a lower probability of false alarm can be achieved due to the lower variance of the blue histogram. The other way round, for the same probability of false alarm, a lower threshold level can be used, which increases the detection rate.

7.2.4 Angle Estimation

In Subsection 7.2.1 the measurement principle of antenna arrays has been introduced briefly. In general, the angle estimation is based on the measured phase offset ϕ_n between different antenna positions (cf. Figure 7.8).

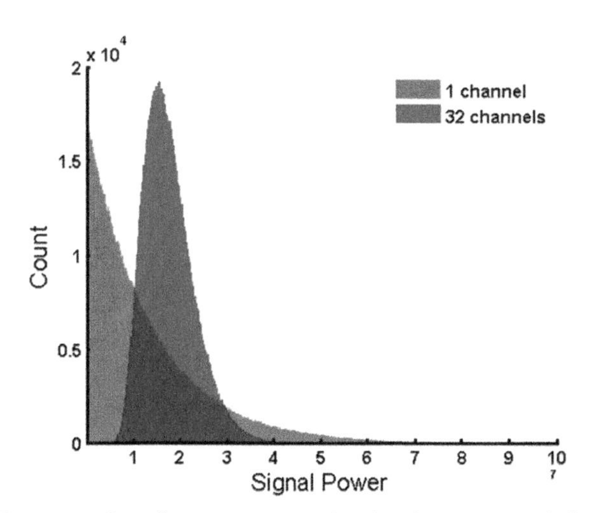

Figure 7.7 Histogram of a noise measurement showing the chi-squared distribution before and after NCI.

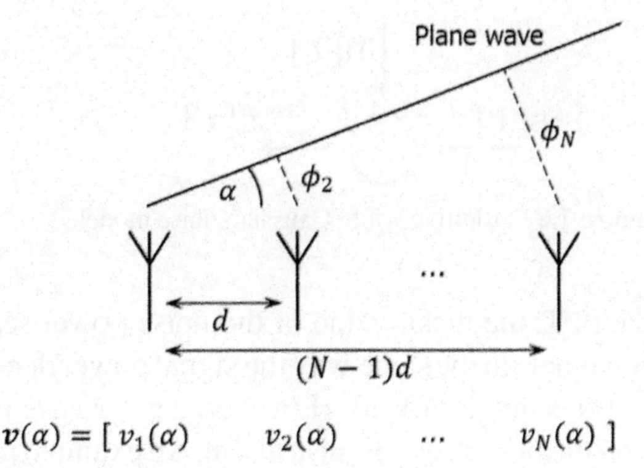

Figure 7.8 Uniform linear antenna array with spacing d and resulting steering vector $\boldsymbol{v}(\alpha)$.

Since the antenna positions are known, a conclusion may be drawn on the direction of arrival. For this purpose, the introduction of a steering vector $\boldsymbol{v}(\alpha)$ can be useful. This vector contains the expected phase offsets, equivalent to an ideal incident signal from a certain angle α:

$$\boldsymbol{v}(\alpha) = \left[e^{j\phi_1(\alpha)} \quad e^{j\phi_2(\alpha)} \quad e^{j\phi_3(\alpha)} \quad \ldots \quad e^{j\phi_N(\alpha)}\right]$$

In the case of a linear array with N elements, the steering vector is simply constructed from the distance d between two antenna elements, the wavelength λ and the incident angle α. The phase of the first element is normalized to zero and the amplitudes are assumed to be all equal one:

$$\boldsymbol{v}(\alpha) = \left[1 \quad e^{j2\pi \cdot d \,\sin\alpha/\lambda} \quad e^{j2\pi \cdot 2d \,\sin\alpha/\lambda} \quad \ldots \quad e^{j2\pi \cdot (N-1)d \,\sin\alpha/\lambda}\right]$$

Similar to the spectral estimation, different classes of algorithms can be identified. Some procedures like the Bartlett beamformer just calculate a weighted sum of the received signal vector $\boldsymbol{x}$. This is done for each possible DOA and results in an angular spectrum:

$$P(\alpha) = \left|\boldsymbol{x}^T \boldsymbol{v}(\alpha)\right|^2$$

The magnitude P represents the correlation between the received signal and the steering vector. A subsequent maximum search extracts the estimated target angle. The separation of two targets is also possible by simply extracting the two largest peaks, however attention has to be paid to the occurrence of sidelobes. Furthermore, the width of the mainlobe determines the separability which is often not satisfactory.

More sophisticated methods to mention are the Capon beamformer, also known as minimum variance estimator, which achieves a better angular separability. Another important class is known as subspace based methods, incorporating MUSIC and ESPRIT as the most prominent examples. Finally, maximum-likelihood estimators exist, which need to know the model order in advance, i.e. the number of targets. However, if the targets have already been separated by different ranges and velocities, the estimation of the model order is feasible because only few targets will be present, in most of the cases only one. A comprehensive overview of existing methods and algorithms is given in [2].

7.3 Hardware Accelerators for MIMO Radar Systems

7.3.1 Basic Structure of a Streaming Hardware Accelerator

Figure 7.9 shows the overview of a hardware-accelerator for high-resolution MIMO radar sensors. Obviously, a high degree of parallelism can be observed, due to the pair wise independence of the receiving channels. Up to the NCI step, each data stream is processed for its own.

The spectral analysis is carried out with the help of a FFT, whose efficient implementation in streaming applications is well understood. A critical step in the design process of this block is the specification of the maximum FFT lengths, as this parameter determines essentially resource usage. Furthermore, when using fixed-point arithmetic, the word length and data scaling behavior can have major effects on performance and efficiency. This aspect is investigated in Subsection 7.3.2.

Regarding the two-dimensional FFT, a concept for data storage and transfer has to be developed. The storage of a complete chirp sequence, i.e. a set of K ramps is required in order to perform the second dimension FFT processing. This dictates mainly the size of the memory, which grows rapidly due to the influence of further key parameters. In general, increasing the resolution in range, in velocity or in the angular domain, also increases the required memory size. It turns out, that this size exceeds rapidly several

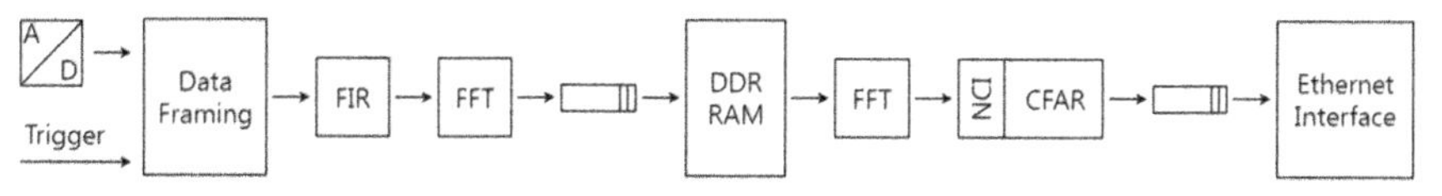

Figure 7.9 Architecture of a streaming hardware accelerator.

MBytes. Thus, the usage of large DRAMs becomes necessary since the size of an on-chip SRAM cache memory is not sufficient anymore. An analysis for different modulation and system parameters can be found in [12].

Regarding the throughput of the memory, the addressing scheme affects heavily the performance in the case of a DRAM. The row opening and closing delays, as well as the read and write transfers can be completely hidden due to the streaming nature of the application. The problem of transforming large two-dimensional matrices with the help of DRAMs has been investigated in [13]. An addressing scheme suitable for the application to chirp-sequence processing has been derived in [14].

Depending on the type of threshold estimation, the calculation can be a simple mean value in the case of CA-CFAR, but it can also become very costly in the case of a sorted list (OS-CFAR). Subsection 7.3.3 presents an efficient architecture based on the rank-only OS-CFAR which avoids a complete sorting of the values inside the window.

7.3.2 Pipelined FFT Accelerator

For streaming applications, pipelined FFT architectures provide a very high throughput. The usage of dedicated hardware accelerators is especially useful for real-time applications, where a high degree of capacity utilization can be achieved. Many different implementation forms have been reported in the past decades. One important parameter is the used butterfly architecture, which can be based on a Radix-2, Radix-4 or Split-Radix decomposition, just to mention a few. In practice, multiple butterflies are cascaded to achieve longer transform lengths. Another important design decision is the use of a single-path vs. multi-path implementation.

A straight forward implementation of the Cooley and Tukey FFT algorithm is shown in Figure 7.10 [15]. It is realized with Radix-2 butterflies which are combined in a decimate-in-frequency (DIF) decomposition. This architecture can process one sample per clock cycle and needs $\log N - 1$ multipliers.

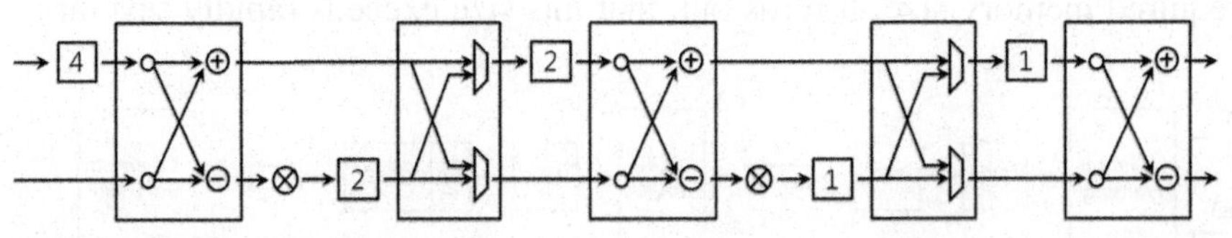

Figure 7.10 Radix-2 FFT implementation based on a multi-path delay commutator (MDC) pipeline.

Furthermore, several buffer memories are required which have the total size $3N/2$.

When analyzing the data flow, it turns out that the butterflies and the multipliers are only used half of the time. Furthermore, only half of the memories store valid data at the same time. Several optimizations have been proposed in order to increase the utilization of the multipliers and memories. For example when using feedback networks, the efficiency in terms of memory usage can be improved. This class of pipeline architectures is known as single-path delay feedback (SDF) network (cf. Figure 7.11) [16].

When using Radix-4 butterflies, the number of multipliers can be reduced as well, at the cost of more complicated butterflies requiring more dedicated adders.

Another FFT algorithm for pipelined implementations has been proposed by He and Torkelson [17] and is known as Radix-2^2 algorithm. This optimization simplifies the traditional Radix-2 FFT decomposition by considering two butterfly stages at once. When modifying some of the twiddle factors, all multiplications after the first stage can be omitted or rather transformed into a trivial multiplication by $\pm$j. Adopting this modification to the presented Radix-2 SDF architecture, half of the multipliers can be saved. Table 7.1 compares different implementations.

In the case of multiple parallel data streams, the utilization of the complex adders and multipliers can be further increased to 100% by using a modified MDC architecture with a proper scheduling of the different data streams [18].

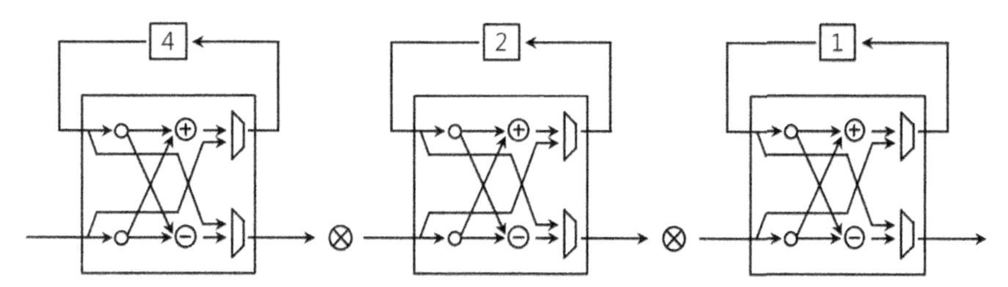

Figure 7.11 Radix-2 FFT implementation based on a SDF pipeline.

Table 7.1 Resource usage of different pipelined FFT implementations [17]

	No. of Multipliers	No. of Adders	Memory Size
Radix-2 MDC	$2(\log_4 N - 1)$	$4\log_4 N$	$3N/2 - 1$
Radix-2 SDF	$2(\log_4 N - 1)$	$4\log_4 N$	$N - 1$
Radix-4 SDF	$\log_4 N - 1$	$8\log_4 N$	$N - 1$
Radix-2^2 SDF	$\log_4 N - 1$	$4\log_4 N$	$N - 1$

In the case of MIMO systems this approach outperforms the Radix-2^2 SDF implementations which seem to be superior in single channel applications.

Even though not optimal in terms of butterfly utilization, a Radix-2 based architecture provided by the Xilinx IP Core is used for the presented MIMO radar system [19]. The principal reason is the faster implementation and integration time. The efficiency in terms of resource usage can be improved in future work.

Fixed-Point Noise

In digital signal processing systems, all computations are carried out with discrete values. The majority of arithmetic units use fixed word lengths which always have a limited accuracy. Consequently some amount of quantization noise is added for each rounding operation. Often floating-point values are used, because they work very well in most environments, regardless of the input signal characteristics. However, if the dynamic range of the input signal is known to a certain extent, fixed-point arithmetic can considerably reduce the resource usage. Many FFT accelerators use integer operations and various models for the engendered quantization noise have been developed.

The quantization noise due to truncation or rounding after a multiplication is often modeled as additive white noise source with a uniform distribution. Even though not accurate under all circumstances, this model is appropriate if the input signal has a sufficiently large bandwidth and amplitude [20]. It can thus be applied to a radar system, due to the wide bandwidth background noise, which is always visible.

The quantization noise variance σ^2 in the case of a uniform distribution can be derived for a simple truncation [21]. The least significant bit (LSB) after the truncation is denoted by $q = 2^{-b}$, where b is the resulting integer word length and k the number of truncated bits:

$$\sigma^2 = \frac{q^2}{12}(1 - 2^{-2k})$$

During the computation of the FFT, the variables grow with each butterfly stage, resulting from the addition inside the butterflies. The complex multiplication does not scale up the intermediate values, because they perform just a rotation in the complex plane and the twiddle factors are all normalized. Thus, the resulting word length of the FFT depends on the input data and grows by 1 bit with each stage. In order to maintain a certain word length, the values can be scaled after each stage at the cost of additional quantization noise.

A complete scaling of the input signal is disadvantageous and engenders an even higher level of quantization noise [22].

Furthermore, a quantization error is introduced after the multiplication, because the resulting word length is cut down by half and also the twiddle factors are represented with limited accuracy. However, it turns out that the coefficient errors are less severe than the round-off errors if the same word length is used for both the coefficients and signals [22].

The following analysis is based on [22], and only the most severe round-off errors are considered. The used noise model applies to a Radix-2 decimation-in-frequency butterfly, which is used by the presented system. Furthermore, the signals are not scaled directly after the addition, but only after the multiplication. Therefore, only one noise source is present for each butterfly output. For the sake of simplicity, the error variance for both outputs is considered equal, even though only one output is the result of a multiplication. This approximation acts as an upper bound because the real output variance after the addition and the truncation will be slightly lower.

The variance of the quantization error σ_e^2 after the multiplication is derived by decomposing the complex operation into four real multiplications, each truncated individually. In this case, the uniform noise model is applied and the number of truncated bits k is assumed to be sufficiently large:

$$\sigma_e^2 = 4\frac{q^2}{12} = \frac{q^2}{3}$$

The total output variance is then calculated by adding all error variances contributing to the respective output. When observing the butterfly graph, a tree-like structure leads to each output, incorporating $N-1$ butterfly nodes. However, if the signal is scaled after each stage, the accumulated noise decreases just as well. In this case, the total noise variance σ_N^2 equals to:

$$\sigma_N^2 = \sigma_e^2 + 2\frac{\sigma_e^2}{4} + 4\frac{\sigma_e^2}{16} + \cdots + \frac{N}{2}\frac{\sigma_e^2}{(N/2)^2} =$$

$$\sigma_N^2 = \left(1 + \frac{1}{2} + \frac{1}{4} + \cdots + \frac{1}{N/2}\right)\sigma_e^2 \approx 2\sigma_e^2$$

Remarkably, the total noise variance is independent of the length of the FFT. However, when examining the signal-to-noise ratio (SNR) at the output, it turns out that the SNR is decreasing for longer FFT lengths, because the output is a scaled version of the FFT. Considering a random input signal,

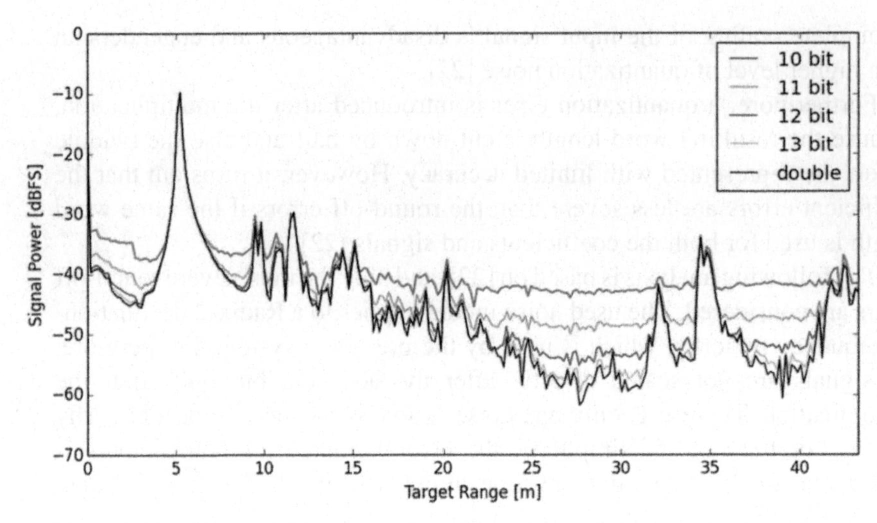

Figure 7.12 Effects of different word lengths on the amount of quantization noise.

with all values i.i.d. and a variance σ_s^2, then the variance for each output of the FFT is scaled by $\frac{1}{N^2}$:

$$\sigma_{s,fft}^2 = \frac{1}{N^2}(N\sigma_s^2) = \frac{\sigma_s^2}{N}$$

Composing the signal-to-noise ratio at the output leads to the expected result:

$$\text{SNR} = \frac{\sigma_{s,\ fft}^2}{\sigma_N^2} = \frac{\sigma_s^2}{2\sigma_e^2 N} = \frac{3\sigma_s^2}{2Nq^2}$$

Consequently, if the FFT length N is doubled, the word length has to be increased by half a bit also in order to maintain a constant signal-to-quantization-noise ratio (SQNR). To illustrate the influence of the word length, an exemplary radar measurement, processed with a scaled fixed-point FFT is shown in Figure 7.12.

Different word lengths have been used in order to illustrate the effect of the introduced quantization noise. The FFT is implemented in a Radix-2 DIF decomposition. The values are rounded and scaled after each stage. The black curve has been processed with double precision floating-point and acts as reference.

It can be observed that the fixed point versions lie all above the reference. The reason is that the quantization noise power is added to the signal and the amount of quantization noise should be the lowest for the floating point version. Furthermore, it can be observed that the noise floor increases significantly in regions with low signal power. The difference for 1bit word length is about 6dB, which correlates with the derived noise model in the case of a truncation or rounding operation. In regions with more signal power, for instance around 5m target range, the quantization noise effect is less severe, due to the higher SQNR.

For a radar system application, it should be ensured that the added quantization noise does not deteriorate the total signal-to-noise ratio. The SNR is a key parameter for reliable target detection. Noise components arising from fixed-point computations should be clearly below the system noise floor in any case. It is important to consider the processing gain when designing an optimal word length, because the noise level drops for larger FFT lengths. Thus, the maximum possible FFT length can be considered as worst-case scenario when designing the word length of the FFT.

7.3.3 Rank-Only OS-CFAR Accelerator

The CFAR processing step requires the use of a local window for threshold calculation. For a streaming application, a sliding window exploits the locality of the data and can be used easily without excessive memory transfers. It is implemented with the help of a shift register. Current FPGA devices offer several different building blocks for this purpose, namely Block RAMs, lookup tables (LUTs) and ordinary flip-flops. For the presented OS-CFAR architecture all signal values inside the window need to be accessed at once. Hence, a data tap is required at each position of the shift register and solely flip-flops can be used for its realization.

As described in subsection 7.2.3, the rank-only detection step depends on N comparisons, a binary sum and a comparison for the decision. Each register of the sliding window is routed to a dedicated comparator, whose second input is fed by the CUT with a threshold value applied. The comparison result is routed to a binary adder with N inputs. Several LUTs are cascaded for this step, which can impose an upper limit to the clock frequency. In order to maximize performance, it is implemented in two steps, i.e. the lower and the upper half of the window is summed up separately before the final rank is computed.

The described architecture has been implemented on a Virtex-7 FPGA and the engendered resource usage has been analyzed. For window sizes up to 128, an operating frequency of 250 MHz could be achieved by this

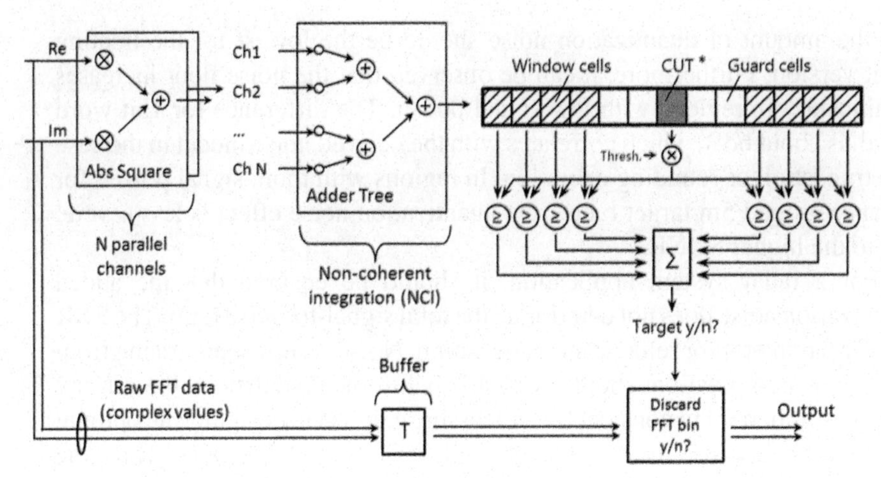

Figure 7.13 Architecture of the rank-only OS-CFAR accelerator.

implementation. The LUT usage depending on the number of channels is depicted in Figure 7.14.

As expected the CFAR-processing part (greenish blue color in Figure 7.14) is practically independent from the number of channels, because the NCI step is performed in advance. The NCI step by itself scales approximately with $\log N$, which is a result of the used tree structure. For a number of channels above 32, the raw data buffer which compensates the pipeline delay consumes more LUTs than the CFAR processing part. It grows linearly with the number of channels and is thus the dominating part for large channel numbers. The

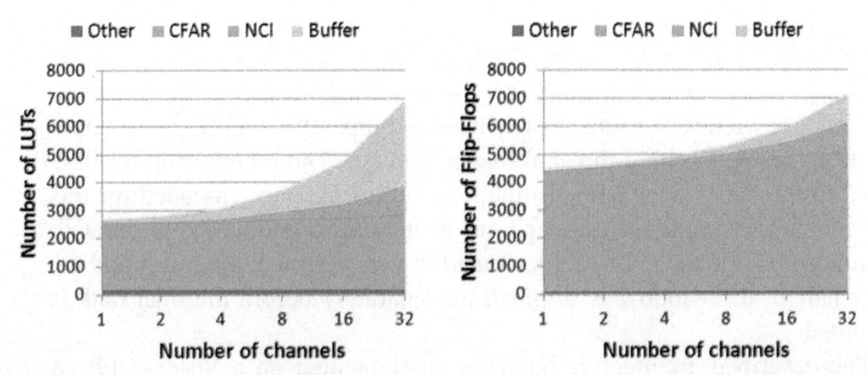

Figure 7.14 Resource usage against number of channels for a constant window size (128 cells).

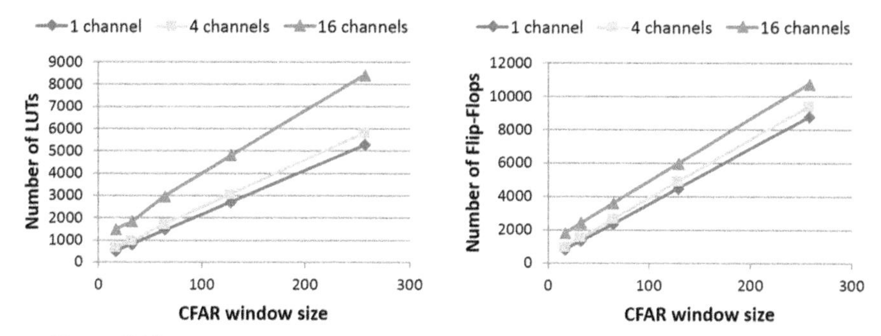

Figure 7.15 Resource usage against window size for different number of channels.

usage of a dedicated Block RAM can be considered if the number of LUTs is scarce.

The scaling behavior in relation to the window size turns out to be nearly linear (cf. Figure 7.15). It is clearly dominated by the N comparators as well as the data buffer equalizing the pipeline delay. The number of channels has a much lower effect on LUT resource usage as the window size. For instance, the resource usage is within the same order of magnitude when comparing one and 32 channels. The architecture can be considered as very efficient for large channel numbers and is thus suitable for MIMO systems. It can be concluded that the usage of NCI before the actual CFAR processing is beneficial in two ways. It improves detection performance and reduces resource requirements at the same time.

7.4 Conclusion

A data processing architecture for future automotive MIMO radar systems has been presented in this chapter. Beside the algorithmic background information, a focus has been set on the target detection with the help of CFAR processing. Attention has been paid to real-time requirements as well as resource usage. The step between the target detection and the subsequent angular processing could be identified as a good data interface between different processing units, each optimized for different requirements on control flow complexity and data throughput.

Furthermore, a FPGA based implementation of the raw data preprocessing chain has been presented and investigated. As crucial points in the design procedure, several parameters could be identified. Especially, the maximum length of the FFTs and the expected dynamic range of the signals determine basically the resource usage in terms of logic elements and memory size. These parameters have a strong dependency on the used modulation waveform, which is why the design of the signal processing architecture has to be integrated into the overall radar system design process. With the help of model-based design space exploration methods, the estimation of resource requirements is feasible, even in an early development stage. The derivation of appropriate models from the realized hardware implementation will be part of future work.

The used design methodology which evolved from the DESERVE project turned out to be very efficient in terms of performance and development time. The usage of heterogeneous platforms, even in an early prototype system, made it possible to handle the tremendous amount of data in real-time. Thanks to the integration with established tools like ADTF and Matlab, the system is ready to be integrated into a test vehicle with a multiplicity of sensors devices. Finally, the early availability of such high resolution automotive radar sensors can be an important step on the way towards automated driving.

References

[1] V. Winkler. “Range Doppler detection for automotive FMCW radars.” *IEEE 37th European Microwave Conference (EuMC),* Munich, Germany, 2007.

[2] H. L. van Trees. “Optimum Array Processing (Part IV of Detection, Estimation, and Modulation Theory)” *John Wiley & Sons,* 2004.

[3] U. Nickel. “Angular superresolution with phased array radar: a review of algorithms and operational constraints.” *IEE Proceedings F: Communications, Radar and Signal Processing* 134.1 (1987): 53–59.

[4] D. Kellner, M. Barjenbruch, J. Klappstein, J. Dickmann and K. Dietmayer. “Wheel extraction based on micro doppler distribution using high-resolution radar.” *IEEE MTT-S International Conference on Microwaves for Intelligent Mobility (ICMIM),* Heidelberg, Germany, 2015.

[5] M. Bouchard, D. Gingras, Y. De Villers and D. Potvin. “High resolution spectrum estimation of FMCW radar signals.” *IEEE 7th SP Workshop on Statistical Signal and Array Processing,* Québec, Canada, 1994.

[6] M. A. Abou-Khousa, D. L. Simms, S. Kharkovsky and R. Zoughi. “High-resolution short-range wideband FMCW radar measurements based on MUSIC algorithm.” *IEEE Instrumentation and Measurement Technology Conference (I2MTC),* Singapore, 2009.

[7] J. B. Billingsley et al. "Statistical analyses of measured radar ground clutter data." *IEEE Transactions on Aerospace and Electronic Systems* 35.2 (1999): 579–593.

[8] B. Magaz and M. L. Bencheikh. "An efficient FPGA implementation of the OS-CFAR processor." *IEEE International Radar Symposium (IRS),* Wroclaw, Poland, 2008.

[9] R. Perez-Andrade, R. Cumplido, C. Feregrino-Uribe and F. M. Del Campo. "A versatile hardware architecture for a constant false alarm rate processor based on a linear insertion sorter." *Digital Signal Processing* 20.6 (2010): 1733–1747.

[10] M. R. Bales, T. Benson, R. Dickerson, D. Campbell, R. Hersey and E. Culpepper. "Real-time implementations of ordered-statistic CFAR." *IEEE Radar Conference (RadarCon),* Atlanta, USA, 2012.

[11] M. A. Richards. "The discrete-time Fourier transform and discrete Fourier transform of windowed stationary white noise." *Georgia Institute of Technology, Tech. Rep,* 2007.

[12] F. Meinl, M. Kunert and H. Blume. "Massively parallel signal processing challenges within a driver assistant prototype framework: first case study results with a novel MIMO-radar." *IEEE International Conference on Embedded Computer Systems: Architectures, Modeling, and Simulation (SAMOS),* Samos, Greece, 2014.

[13] S. Langemeyer, P. Pirsch and H. Blume. "Using SDRAMs for two-dimensional accesses of long $2^n \times 2^m$-point FFTs and transposing." *IEEE International Conference on Embedded Computer Systems: Architectures, Modeling, and Simulation (SAMOS),* Samos, Greece, 2011.

[14] F. Meinl, E. Schubert, M. Kunert and H. Blume. "Realtime FPGA-based processing unit for a high-resolution automotive MIMO radar platform." *IEEE 12th European Radar Conference (EuRAD),* Paris, France, 2015.

[15] L. R. Rabiner and B. Gold. "Theory and application of digital signal processing." *Prentice-Hall, Inc.,* 1975.

[16] E. H. Wold and A. M. Despain. "Pipeline and parallel-pipeline FFT processors for VLSI implementations." *IEEE Transactions on Computers* 100.5 (1984): 414–426.

[17] S. He and M. Torkelson. "A new approach to pipeline FFT processor." *IEEE 10th International Parallel Processing Symposium (IPPS),* Honolulu, USA, 1996.

[18] K. J. Yang, S. H. Tsai, and G. C. H. Chuang. "MDC FFT/IFFT processor with variable length for MIMO-OFDM systems." *IEEE Transactions on Very Large Scale Integration (VLSI) Systems* 21.4 (2013): 720–731.

[19] Xilinx Inc. "LogiCORE IP FFT." *PG109 v9.0,* October 2014.

[20] C. W. Barnes, B. N. Tran and S. H. Leung. "On the statistics of fixed-point roundoff error." *IEEE Transactions on Acoustics, Speech and Signal Processing* 33.3 (1985): 595–606.

[21] D. Menard, D. Novo, R. Rocher, F. Catthoor and O. Sentieys. "Quantization mode opportunities in fixed-point system design." *IEEE 18th European Signal Processing Conference (EUSIPCO),* Aalborg, Denmark, 2010.

[22] C. J. Weinstein. "Quantization effects in digital filters." *Lincoln Laboratory, Massachusetts Institute of Technology, Tech. Rep. No. TR-468,* 1969.

8

Self-Calibration of Wide Baseline Stereo Camera Systems for Automotive Applications

Nico Mentzer[1], Guillermo Payá Vayá[1], Holger Blume[1], Nora von Egloffstein[2] and Lars Krüger[2]

[1]Institute of Microelectronic Systems, Leibniz Universität Hannover, Hannover, Germany
[2]Daimler AG, Vision Enhancement, Ulm, Germany

8.1 Introduction

Many car accidents involving vulnerable road users (e.g., pedestrians or cyclists) occur on rural roads after dark, when the driver's visibility is restricted. Thus, the main objective of an augmented night vision is to assist the driver, when driving on side roads (e.g., highways, country roads, or rural roads) with poor or restricted visibility by alerting the driver to potential obstacles ahead.

One possible augmentation of driver vision is to highlight potential obstacles, hazards or vulnerable road users in the live video of the road ahead. A classification of image content is mandatory for this application. As the augmentation enables the driver to grasp the situation quickly, the distance to the detected object has to be calculated by stereo vision to ensure accuracy and speed of assessment.

As the range of distance resolution increases with the baseline of a stereo system, a wide baseline stereo system is necessary to facilitate the augmentation of objects in the desired range. Such a wide-baseline stereo system is sometimes not practicable when rigidly coupled, therefore cameras are mounted individually, e.g., to the windshield. Physically separated cameras increase the camera baseline, however a moving car causes multiple vibration sources [1] which misalign the images of the separated cameras. Therefore, online camera calibration is indispensable for further image processing. This online camera calibration covers the reconstruction of extrinsic camera parameters, which rely on a sparse pixel correspondence list from the two camera images. The general overview of the algorithmic flow is depicted in Figure 8.1. This chapter will focus on the search for sparse pixel correspondences and extraction of camera calibration parameters.

The remaining chapter is set up as follows. Section 8.1 gives an introduction to the self-calibration of wide baseline stereo cameras. After a review of the considered algorithms in Section 8.2, Section 8.3 details the class of image feature detectors and extractors. Section 8.4 highlights the matching of image features. An in-depth description of the bundle adjustment for the camera calibration is given in Section 8.5. In Section 8.6, selected application-specific aspects regarding the algorithmic parameterization are presented. Section 8.7 focuses on algorithmic-specific and hardware-specific implementation details and gives an overview of existing implementations for the extraction of image features.

8.1.1 Extraction of Image Features

Image feature extraction consists of two steps: the detection of image features and the generation of the descriptor for those feature points, which results in a unique signature as a representation for the detected feature points.

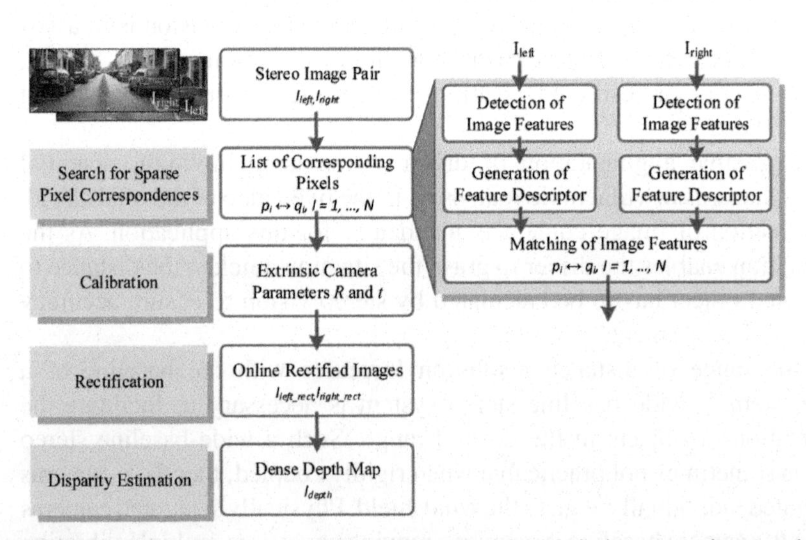

Figure 8.1 Algorithmic overview. Input of the processing chain is a stereo image pair, in which sparse pixel correspondences are extracted for online camera calibration. After the calibration, rectification is performed as a preprocessing step for disparity estimation.

The image feature detection generates a list of distinctive invariant points in images for the feature localization. Especially for camera calibration, a high accuracy of localization is required [2] in order to ensure a correct functionality of following algorithmic steps, e.g., the rectification of stereo image pairs. Due to the similarity between the views of the scene, a rotation invariance or scale invariance of the feature descriptors supports stability of the matches. This is however, not mandatory, because characteristic points in image pairs of the used stereo camera configuration rarely change their rotation or scale abruptly from left to right stereo image.

In recent years, three different principles for **feature detection** have proven employable. *Corner* or *edge detectors* extract characteristic corners or edges in an image, which are defined by large gradient changes of image intensities. So called *blob detectors* determine pixel positions, for which a circular local neighborhood is approximately constant or similar for a defined image property [3]. Furthermore, *affine invariant detectors* have been adapted to be invariant to affine transformations, which are approximations to perspective distortions in order to achieve invariance to large changes in viewpoint [4]. The detected features of the exemplary SIFT-feature detector are shown in Figure 8.2.

Figure 8.2 Left (*top*) and right (*bottom*) image from a stereo camera system showing detected SIFT-image features. Detected feature points of the left/right image are displayed in red/green, matches are displayed in blue. Scale and rotation of the SIFT-features are illustrated by the circle properties.

The descriptor of an image feature characterizes the detected feature point. Ideally, a **feature descriptor** of a world point is unique when compared to other descriptors, but identical for the same world point in different views [5]. Two representations for descriptors have been established in recent years. So called *histogram-based or distribution-based descriptors* represent the local neighborhood of a feature point by histograms of local image properties like pixel intensities, color, texture, edges etc. [3]. Furthermore, *binary descriptors* represent a local pixel region by storing the binary result of predetermined pixel-level intensity comparisons [6]. In contrast to distribution-based descriptors, binary descriptors contain a more compact representation of the image patch around a feature.

In general, extracted image features have to cope with various influences. Firstly, there are disruptive effects related to the image quality, e.g., image compression, image noise, image blur due to zoom or exposure. Secondly, there are influences resulting from the content of the stereo image pair, e.g., illumination, difficult viewpoint conditions or occlusions, background clutter and general content changes, perspective changes or changes in the view point of planar and non-planar geometry [6, 7]. Finally, application specific factors as scale and rotation of objects impact the algorithmic results dealing with image features. Thus, extracted image features have be invariant to as many disturbing influences of the named categories as possible.

The large variety of image feature detectors and descriptors clearly show the manifold approaches to defining and describing characteristic points in images. As S. Gauglitz mentioned before in [5], “there is no clear-cut definition as to what makes a point interesting. Detection of such points is only an intermediate step in any application”. There is no general answer for the question, which detector or descriptor is performing the best. Therefore, as J. Shi and C. Tomasi postulated in 1994, “the right features are exactly those that make the tracker work best” [8]. Consequently, “any set of feature points is acceptable, but the result ought to be consistent, e.g., in images that show the same scene, the algorithm should detect the same points.” [5]. In other words, for each application, the best performing combination of image feature detector and extractor has to be found. Furthermore, application-specific conditions (here: high localization accuracy with low requirements to scale and rotation invariance) aggravate the possibilities of algorithmic combinations.

A survey of existing image feature detectors and descriptors will be given in Section 8.2. A more detailed presentation of an exemplary feature detector and descriptor called SIFT (Scale-Invariant Feature Transform) [9], which shows good results in this application, will be given in Section 8.3.

8.1.2 Matching of Image Features

Matching image features results in a list of pixel correspondences between the left and right input image of the stereo image pair. The main challenge is on the one hand to find as many corresponding pixels as possible while avoiding wrong pixel assignments on the other, even if there are several similar regions in both input images. The assignment of image features to pixel correspondences is based on feature descriptors, which are used to find the maximum similarity between the extracted image features. Depending on the representation of the features (histogram-based or binary descriptor), the similarity is computed by various vector norms for the distance of two matching candidates or the Hamming distance. Furthermore, different matching methods have a significant impact on the resulting correspondence lists [3].

In the case of global feature matching methods $\mathrm{f} : \widetilde{X} \rightarrow \widetilde{Y}$, two feature points $\overrightarrow{x} \in \widetilde{X}$ and $\overrightarrow{y} \in \widetilde{Y}$ are assigned by local similarity, which is determined by the related descriptors $\overrightarrow{d_x}$ and $\overrightarrow{d_y}$. For each descriptor in set Y, there is a corresponding descriptor in set X with a minimal error criterion. After the assignment of feature points, the correspondences are filtered by this error criterion in order to avoid false correspondences, e.g., feature points which are not detectable in both images because of occlusions in one image. Varying matching methods differ in the error criterion for the evaluation of feature similarity and the search algorithm during the matching step.

8.1.3 Extrinsic Online Self-Calibration

Common stereo algorithms for disparity estimation (e.g., [10]) rely on exact knowledge about the intrinsic (e.g., focal length) and extrinsic camera parameters (the transformation between two cameras). Calibration errors lead to erroneous reconstruction values. The camera parameters enable the rectification, which is the projection of the camera images to a common image plane and they form the basis for further processing.

The intrinsic parameters may be assumed to be constant and identified using an offline calibration procedure (e.g., [11]). As the cameras are not rigidly coupled here, the extrinsic parameters vary due to vibrations in the car and are assumed to change rapidly from frame to frame. Thus, a one-time offline calibration procedure does not suffice to meet the accuracy requirements of stereo processing. Thus, an online calibration procedure is necessary. While driving the use of calibration targets with known geometry is difficult. Therefore, a self-calibration mechanism is needed.

The idea behind online self-calibration procedures is to estimate the camera parameters based on what is perceived in both cameras. So a preprocessing step to the calibration is a one-to-one identification of scene points visible in both camera images, e.g., a list of sparse pixel correspondences of the stereo camera images.

8.2 Algorithmic Overview

Many approaches have been proposed in recent years for the extraction and matching of image features and for the feature-based camera self-calibration. In the following section, selected aspects for each algorithmic step are reviewed separately.

8.2.1 Survey of Image Features Extraction

The process of extracting image features is split into two algorithmic parts, the detection of feature points and the generation of the feature descriptor. For both steps, a large number of algorithms have been published. In this section, typical examples of each algorithmic step are presented.

8.2.1.1 Detection of features

Which properties of distinctive image points are mandatory for a satisfactory matching of image features depend on the finale application. There is no clear definition as to which extraction strategy is best as it only needs to provide sufficient algorithmic performance during retrieval in the same scene on image sequences from different viewpoints. Therefore, what is *characteristic* for highly distinctive points in images is an application-specific approach, which has led to four basic methods for extracting retrievable points in images.

Edge detection

Edges are stable features, which are detectable over a range of viewpoints and illumination changes [12]. An edge, e.g., the border of an object, is defined by discontinuities in pixel intensities in a single image dimension (see Figure 8.3(b)). Thus, the Canny detector [13] determines the gradient of the input image with the Sobel operator and by evaluating magnitude and orientation of the gradients, the edge's direction and its strength are extractable.

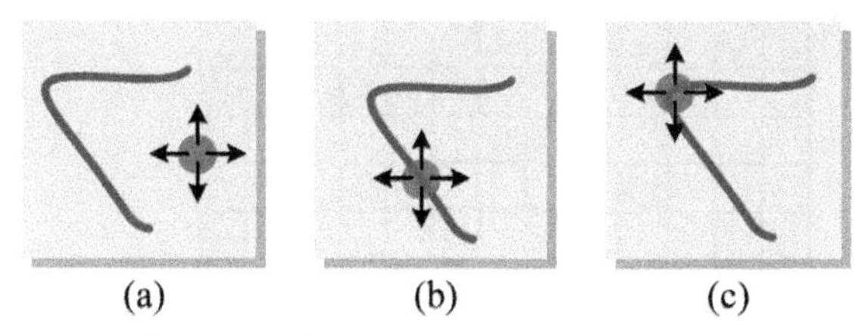

Figure 8.3 Detection of edges and corners by image gradients. The blue circle shows a possible feature point, surrounded by a local neighborhood. (a) Low image gradients in two spatial directions represent texture free image areas. (b) A high image gradient in one spatial direction indicates a possible edge, (c) in two spatial directions a possible corner.

Gradient and direction are used in a non-maximum suppression in order to suppress equivocal edges in the local neighborhood of a possible edge.

The drawback of this method is the equivocalness of the detected feature points. As depicted in Figure 8.3(b), it is not distinct which detected points are corresponding on the edge while matching two detected feature points and therefore, it will lead to incorrect pixel correspondences.

Corner detection

Corners are defined as intersections of edges or as pixel continuities in two or more image directions (see Figure 8.3(c)). In addition to simple corners, line endings and cropped intensity changes are detected using this type of detector.

One early corner detector is the Harris corner detector [14] (1988), which approximates the sum of squared differences of two image patches in order to detect a difference in image intensities. The approximation results in the second moment matrix, which represents the dominant directions of a local neighborhood in the gradient image. With this approach it is not only possible to detect corners, but edges as well.

To avoid such costly filters, a detector has been presented that does not rely on discrete image derivatives, but on the number of intensity differences between pixels [5], which are located on a Bresenham circle (see Figure 8.4). Rosten [15] sped up this process by reducing the number of pixel tests with machine learning techniques to find the fastest sequence of pixel comparisons for rejecting a wrong corner candidate.

The matching of detected corners in different images of the same scene provides correct pixel correspondences as long as the detected corners belong to objects of the same size. A corresponding corner is just detectable in different images, if the regions for describing the corners have similar dimensions

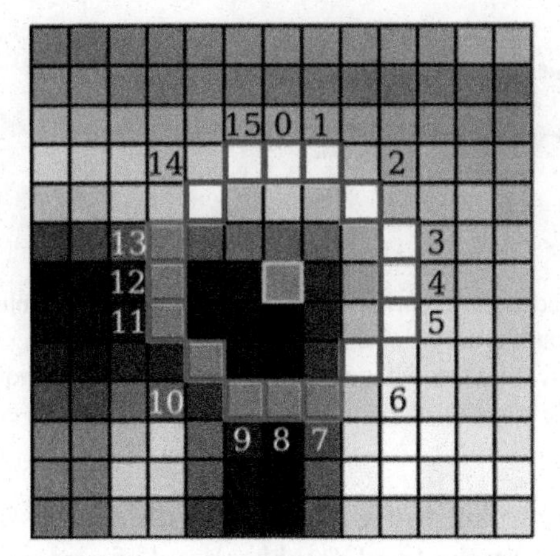

Figure 8.4 Intensity comparisons of pixel, which are located on a Bresenham Circle. The central pixel is determined as a corner if a certain number of continuous pixel intensities is brighter or darker than the central pixel. This is combined with an adoptable threshold to avoid instabilities.

(see Figure 8.5, red circle), which is dependent on the object size. To overcome this problem, repeated image scaling is a possibility or an object size dependent adjustment of the region for the descriptor generation.

Blob detection

A blob is a region of connected pixels, which share a common image property, e.g., pixel intensities, and therefore stand out from surrounding regions. By formulating image properties as a function of pixel positions, local maxima and minima of the function are determinable.

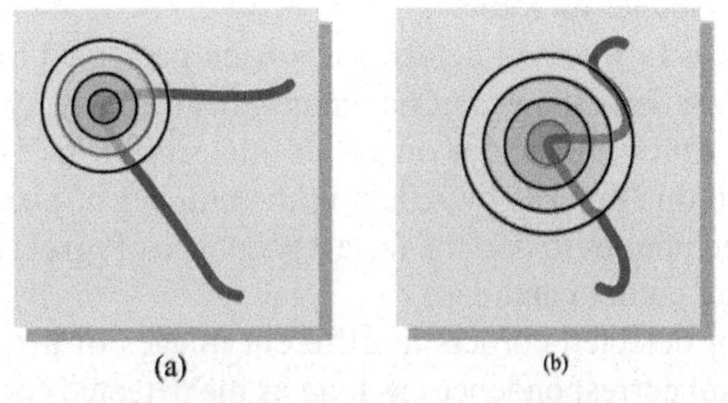

Figure 8.5 Detection of corners of different image scales. With strongly different object sizes in the image, a corresponding corner is not detectable (red circle), but by a repeated image scaling.

Figure 8.6 Blob detector. The detected blobs are displayed as red circles. The blob's size is displayed as the diameter of the circle.

It has been shown, that the Laplacian of Gaussian (LoG) [16] has a strong response to dark and bright image regions, which are detectable as blobs. The response is highly dependent on the size of the filter kernel used (see Figure 8.6).

Affine-invariant interest point detection

Images features based on a blob detector hardly match for large scale or viewpoint changes [4], because circular image patches for blob feature extraction will lead to large distance measures for blob feature matching due to less covering of the circular regions (see Figure 8.7). By applying circular image patches, the used image information is too different to ensure stable pixel correspondences for large viewpoint changes. Therefore, Mikolajczyk [7] extends blob detectors to affine invariance by estimating the affine shape of a local neighborhood. For affine transformations, the scale of an image region changes differently in each direction, which leads to differing local regions for the blob detection and therefore to differing localization or to mistaken detections.

Figure 8.7 Blob detection based on circular image region for a scene with a large viewpoint change. The region on which the blob feature extraction is based only partially covers the corresponding region and thus, will lead to non-matching image features.

In order to deal with affine transformation, Mikolajczyk [7] replaces the blob detection scales, which are equal in all directions, by affine detection scales, which vary independently in orthogonal directions. Hereby, the circular point neighborhood is replaced by an ellipse, which is determined by the second moment matrix. With the affine normalization, the ellipse is normalized to a circle again and a blob is detectable within the transformed image patch (see Figure 8.8).

Since the four presented methods provide large differences in quantity and quality for detected interest points, a suitable algorithm has to be chosen with regards to the application.

In 3D reconstruction, precise localization of interest points is one major aspect [4], therefore a sub-pixel accuracy for feature detection is mandatory. Self-occlusion occurs very frequently in real world scenes and typically many interest points are found near occlusion boundaries. Accurate positioning of features is imperative. As has been shown in many publications, center-oriented detectors (e.g., LoG, DoG or CenSurE) [5], provide a higher and more stable repetition rate than corner or edge detectors. Furthermore, affine-invariant interest point detectors have been adapted to be robust to large changes in viewpoint [4], which is of minor importance even for reliable image feature matching for a wider baseline stereo camera system.

Taking into account the algorithmic robustness of the presented methods for the detection of image features and the high requirements of ADAS (Advanced Driver Assistance Systems), a blob detector is used for the detection of features henceforth. In subsection 8.3.1 the SIFT-detector [9] will be presented in detail as an exemplary blob detector.

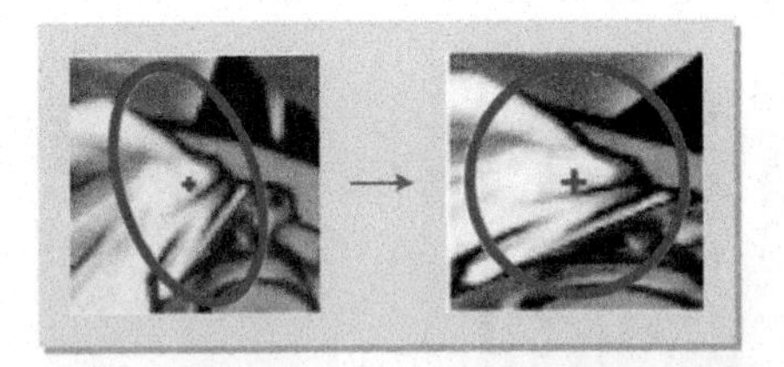
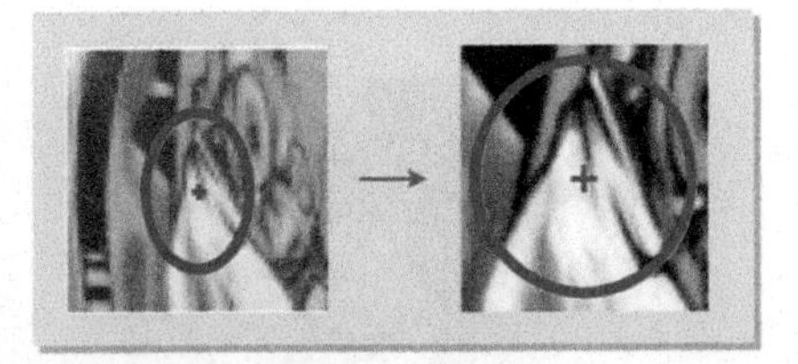

Figure 8.8 Affine-Invariant Interest Point Detection. The circular point neighborhood is replaced with an ellipse in order to achieve independent orthogonal varying detection scales for interest point detection. Before applying a detection algorithm, the local neighborhood is affine normalized, which results in a circular neighborhood and a transformed image patch (from [7]).

8.2.1.2 Description of features

After the detection of interesting points, the descriptor as a unique representation of an image feature has to be generated. In addition to histogram-based descriptors, which are memory greedy, binary descriptors have been established as a more compact representation for image features. In addition, compared to histogram-based descriptors, the distance of two binary descriptors, which is required for feature matching, is faster to match. There are other techniques to describe image features such as image patch correlation or generalized moment invariants [3], however the focus of this section is limited to the two mentioned descriptor types, due to their suitability for the self-calibration of wide baseline stereo camera systems.

Histogram-based descriptors

A simple way to describe a detected blob in a histogram-based manner is the distribution of pixel intensities of the local blob region. Due to the fact that this technique is prone to illumination changes, more complex approaches have been presented (see [3]), e.g., the distribution of gradient locations and orientations in the local blob area instead of the distribution of pixel intensity itself. In the case of the SIFT-descriptor, the coordinates of the descriptor and the gradient orientations are rotated relative to the feature orientation and afterwards, a histogram is generated based on orientation and magnitude of the image gradient [9]. Furthermore, the quantization granularity of gradient locations and orientations leads to a robust descriptor, which is stable to small geometric distortions and small errors in the blob region. Besides multiple techniques for histogram generation, different sampling grids have been introduced (see Figure 8.9). The resulting descriptor is a multidimensional vector with the histogram's bins as components. In the case of SIFT,

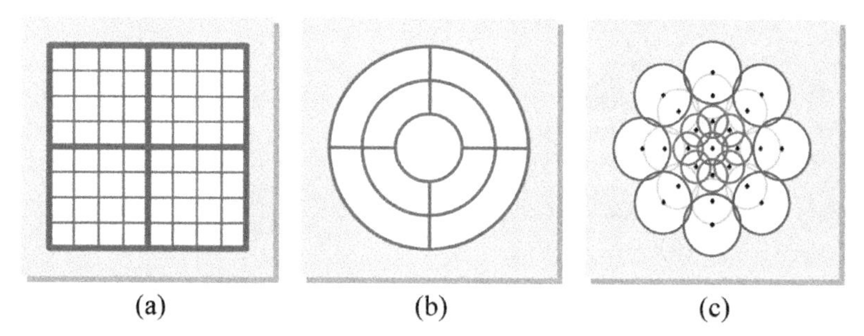

Figure 8.9 Sampling grids for generating different descriptors: (a) SIFT [9], (b) Shape Context [18], (c) DAISY [19].

each vector consists of 128 values of floating point precision. The size of a feature vector is highly dependent on the algorithmic parameters, but nevertheless histogram-based descriptors usually have high memory requirements. Therefore, techniques for a more compact descriptor representation have been developed, e.g., principal component analysis for PCA-SIFT [17].

Binary descriptors

Due to the fact that histogram-based descriptors provide a large complexity [3] and high memory requirements [6], a sped up generation and a more compact representation for feature descriptors is desirable. Therefore, binary descriptors are characterized by sampling patterns and predefined sampling pairs. Sampling patterns define a set of potential sampling locations (Figure 8.10, blue circles), whose image information are optionally smoothed with spatial-dependent filter kernels (e.g., Gaussian smoothing) (Figure 8.10, red circles). A fixed combination of the filtered intensities is selected in advance as descriptor specific sampling pairs (see Figure 8.11, two variations of sampling pairs for the FREAK descriptor).

For each sampling pair, a binary test τ is performed, e.g., (BRIEF [20]):

$$\tau(\mathbf{p}; x, y) := \begin{cases} 1 & if\ I(\mathbf{p}, x) < I(\mathbf{p}, y) \\ 0 & otherwise \end{cases}$$

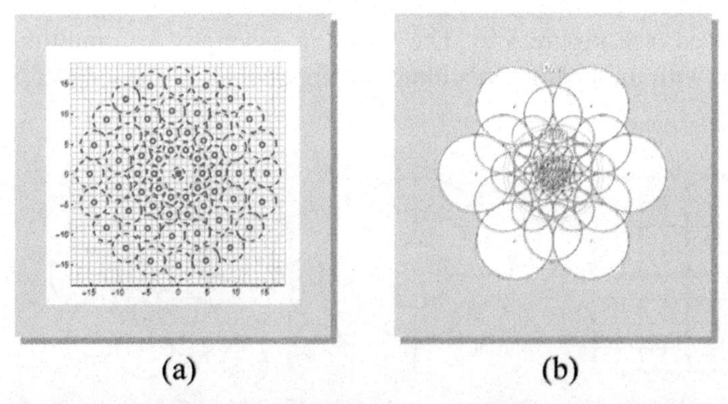

(a) (b)

Figure 8.10 Sampling pattern. (a) BRISK descriptor, (b) FREAK descriptor [21]. Sampling patterns define a set of sampling locations (blue circles), of whose image information is smoothed with spatial-dependent filter kernels (red circles). Out of the sampling pattern the sampling pairs for the binary tests for the descriptor generation are selected.

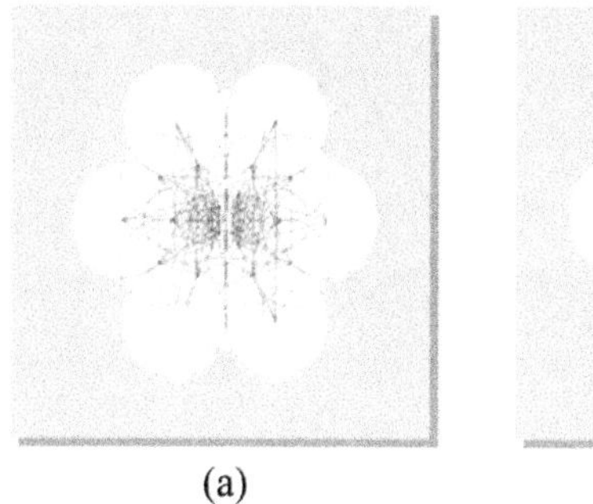
(a)

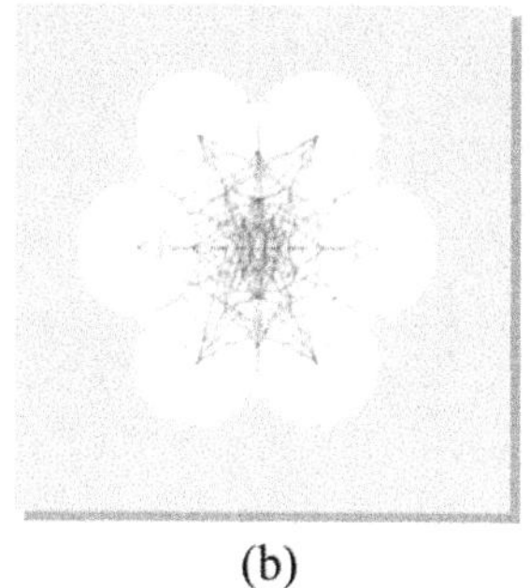
(b)

Figure 8.11 Two variations of sampling pairs of the FREAK descriptor [21]. A fixed combination of sampling locations is selected as descriptor specific sampling pairs, with which the binary tests for the descriptor generation is performed.

where $I(\boldsymbol{p}, x)$ is the pixel intensity in a smoothed image patch $\boldsymbol{p}$ around an image position $x = (u, v)^T$. On a set of n_d precomputed pixel pairs, such binary tests are performed. The resulting descriptor of dimension n_d ensues to

$$\sum_{1 \leq i \leq n_d} 2^{i-1} \tau(\boldsymbol{p}; ; x_i, y_i)$$

Typically, a binary descriptor has a maximal length of 512 Bit.

8.2.1.3 Characteristics of features

Invariances to rotation and scale increase the detection rate of features in similar views of a scene and ensure the distinctiveness of the detected feature points. By assigning a region based main orientation, a feature is rotated by this orientation in order to match it with a corresponding feature from a different orientation. Furthermore, objects often vary in size in different images, which lead to variant image regions for the description of the same feature. To unify the descriptor generation, Lindeberg's [16] scale-space theory is applied.

Rotation invariance of a feature descriptor is achieved by rotating the sampling grid or sampling pattern for the pixel area which is used for the descriptor generation by the main orientation before the descriptor is extracted (see Figure 8.12) or by rotating the descriptor itself. To determine the main orientation, different approaches are available. Rublee et al. [22] use intensity centroids to determine the main orientation of a patch, whereas Leutenegger et al. [23] use the gradient of predefined sampling pairs to rotate the sampling pattern. Further techniques are available in the literature (e.g., [9, 21, 24]).

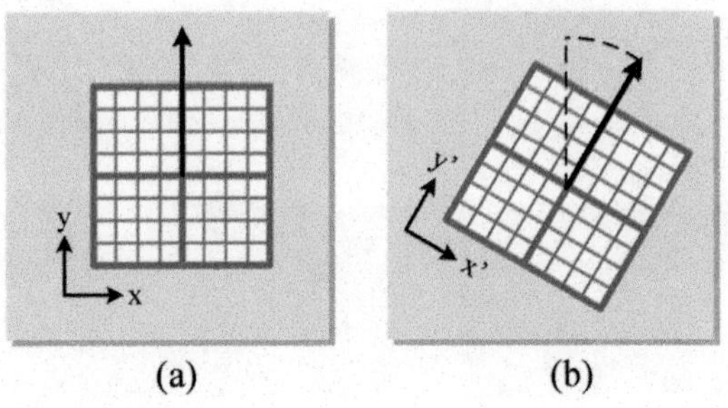

Figure 8.12 Rotation invariance is achieved by rotating the sampling grid by the main orientation before extracting the descriptor.

Scale invariance of image features is attained by applying Lindeberg's [16] scale-space theory for image processing to the input images while detecting image features. The input image is subsampled multiple times to generate different scales of the input image and the detection step is repeated. If the same feature candidates are detected on multiple scales, the candidate on the scale with the highest information content is selected in order to achieve scale-invariance (see Figures 8.13 and 8.14). Lowe (SIFT, [9]) approximates

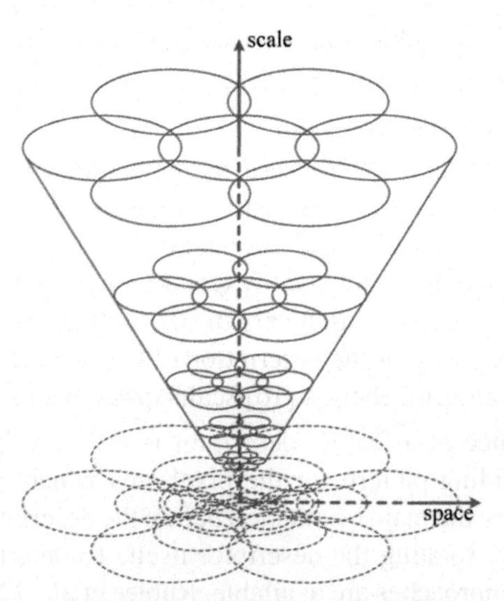

Figure 8.13 Scale-space. An input image is down sampled to achieve multiple scales of the image. On each scale, feature candidates are found, whereas repeated candidates are removed. The scale with the highest information content for the feature candidate is selected as the feature scale (from [16]).

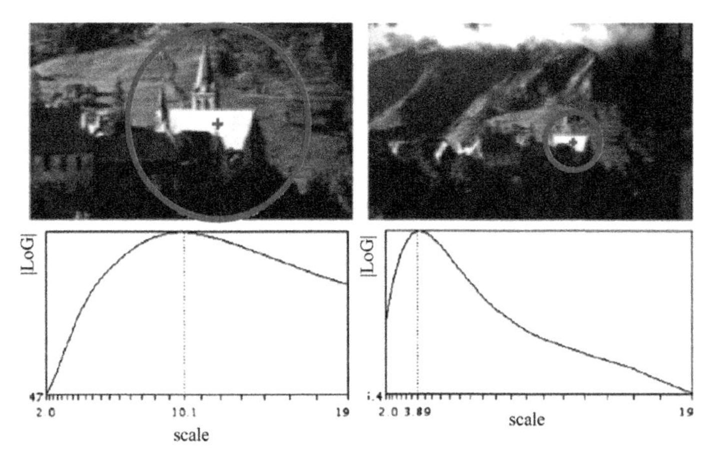

Figure 8.14 Multi-scale approach for blob detection. The same blob with differing scales in two images and the related response (normalized Laplacian of Gaussian) over scales is shown. The scale with the highest information content is chosen as a blob (from [7]).

Lindeberg's LoG scale-space with different Gaussian smoothed images and therefore, the complexity is reduced significantly.

A further approach for scale invariance is the detection and later suppression of feature candidates which are detected on multiple scales, but have the same image position. Those repeated nominations are compensated by a non-maximum suppression [6], which evaluates a predefined cornerness score and selects the most unique feature point.

Image feature detection and description are not completely independent. By choosing a certain feature detector, a specific local neighborhood is used to detect interesting points. This specific local neighborhood has to be also employed to extract the feature descriptor in order to ensure a reliable description of the image patch. Although it seems to be a promising approach, it is not advisable to combine any detector with any descriptor [4]. The following overview (see Tables 8.1 and 8.2) of selected state-of-the-art feature extractors and feature descriptors with references is not intended to be exhaustive, but gives an impression of how many different detectors and extractors are available and therefore combinable. For an appropriate performance, each algorithm requires an application-specific parameterization, which may depend on the previous and following processing step. Thus, this large number of degrees of freedoms results in an algorithmic variety, which is hardly ascertainable.

Table 8.1 Overview of feature detectors

Feature Detector	Year	Comment
SIFT [9]	1999	Scale-Invariant Feature Transform Scale-space based, invariant to scale and rotation
SURF [25]	2008	Speeded Up Robust Features Scale-space based, invariant to scale and rotation
KAZE [24]	2012	Non-linear scale-space based Invariant to scale and rotation
A-KAZE [26]	2013	Accelerated-KAZE Improved KAZE feature detector
BRISK [23]	2011	Binary Robust Invariant Scalable Keypoints Scale-space based, invariant to scale and rotation
FAST [15]	2006	Features from Accelerated Segment Test Segment based corner detector
ORB [22]	2011	Oriented FAST and Rotated BRIEF Advanced from FAST and BRIEF (see descriptors)

Table 8.2 Overview of feature descriptors

Feature Descriptor	Year	Comment
SIFT [9]	1999	Scale-Invariant Feature Transform Histogram-based descriptor
SURF [25]	2008	Speeded Up Robust Features 7 Histogram-based descriptor
KAZE [24]	2012	Non-linear scale-space based Histogram-based descriptor
A-KAZE [26]	2013	Accelerated-KAZE Binary descriptor
BRISK [23]	2011	Binary Robust Invariant Scalable Keypoints Binary descriptor
BRIEF [20]	2012	Binary Robust Independent Elementary Features Binary descriptor
ORB [22]	2011	Oriented FAST and Rotated BRIEF Advanced from FAST and BRIEF (see detectors)
DAISY [19]	2010	Dense Descriptor for Wide Baseline Stereo Matching Histogram-based descriptor
FREAK [21]	2012	Fast Retina Keypoint Binary descriptor

8.2.2 Feature Matching

The final step in finding sparse pixel correspondences is the assignment of the extracted image features in different image set ups, e.g., in time sequentially images for sparse optical flow, in stereo image pairs for feature-based sparse disparity estimation or in image patches for object detection.

As in the case of the previous algorithmic steps, many approaches for descriptor matching have been presented in recent years [3]. In order to determine the similarity of two image features, multiple correspondence measures are available. In addition, various matching methods lead to significant differences in matching results, which influences the resulting pixel correspondence lists and finally, some matching methods require a list search algorithm, for which again different approaches are available. Each aspect will be briefly reviewed in the following subsection.

Correspondence measures for image features

For histogram-based descriptors $\overrightarrow{d} \in \mathbb{R}^l$, which are real-valued vectors of dimension $l \in \mathbb{N}$, multiple vector norms are applicable on matching difference vectors as a similarity measure. The sum norm is defined as the accumulation of the component wise sum of absolute differences:

$$\|\overrightarrow{d}_x - \overrightarrow{d}_y\|_1 = \sum_{i=1}^{l} |d_{x,i} - d_{y,i}|$$

In order to weight large vector difference more than small differences, the Euclidean norm is useable. The norm penalizes large vector differences more than small vector differences by accumulating the component wise sum of squared differences:

$$\|\overrightarrow{d}_x - \overrightarrow{d}_y\|_2 = \sqrt{\sum_{i=1}^{l} |d_{x,i} - d_{y,i}|^2}$$

Since only relative correspondence measures are used for feature matching, the square root is skippable to avoid costly computations.

A further method for evaluating the distance of two vectors is the normalized cross correlation:

$$distance = max_{x \in X}\left(\frac{\sum_{i=1}^{l} d_{x,i} \cdot d_{y,i}}{\sqrt{\sum_{i=1}^{l} d_{x,i}^2} \cdot \sqrt{\sum_{i=1}^{l} d_{y,i}^2}}\right)$$

The correlation yields good results for the matching of image features, but leads to high computational complexity [3] and is therefore rarely used for matching of image features in the field of advanced driver assistance systems.

For binary descriptors, which consist of a bit string of length n, that represent the result of pixel wise test, the correspondence measure is the

Hamming distance, which is the accumulation of the bit wise XOR of the bit strings:

$$ham_{\vec{d}_x, \vec{d}_y} = \sum_{i=1}^{n} (d_{x,i} \oplus d_{y,i})$$

Due to the correspondence measure's simplicity, typically the distance computation of two binary descriptors is noticeably faster than the distance computation of two histogram-based descriptors. Contrary, not every binary descriptor has a comparable quality level as histogram-based descriptors for certain applications. By selecting a specific descriptor type, the implicit trade-off between execution time and descriptor quality has to be taken into account.

Matching methods for image features

The quality of resulting pixel correspondences highly depends on the utilized matching method. Three different methods have been established in the field of feature matching for advanced driver assistance systems (from [3]), which show different behavior in the matching inlier/outlier ratio:

1. **Threshold-Based Matching (TB)**
   Two features match, if the distance between the descriptors is below a predetermined threshold. A feature may have several matches and several of them may be correct.
2. **Nearest-Neighbor-Based Matching (NNB)**
   Two features match, if the descriptor $\vec{d}_y$ is the nearest neighbor to $\vec{d}_x$ and if the distance between the descriptors is below a threshold. A feature only has one match
3. **Nearest-Neighbor Distance Ratio Matching (NNDR)**
   Two features match, if the descriptor $\vec{d}_y$ is the nearest neighbor to $\vec{d}_x$ and if a ratio ε between the first and the second nearest neighbor is below a threshold:

   $$\varepsilon = \frac{\| \vec{d}_x - \vec{d}_y \|_p}{\| \vec{d}_x - \vec{d}_z \|_p}$$

   where p indicates the type of norm. This ratio avoids ambiguous matches in case there are potential matches with a similar distance. Again, a feature has only one match.

The matching quality for both nearest-neighbor approaches are higher than for the TB matching [3], because the probability of a correct match for the nearest neighbor matchings is higher than the TB matching, although the distance between similar descriptors possibly varies significantly. The nearest neighbor matchings select only the best match below the threshold and rejects all others and thus, there are few false matches. In addition, the NNDR matching penalizes descriptors which have many similar matches, e.g., the distance to the nearest neighbor is comparable to the distance of the second nearest neighbor. This leads to further improvement in precision. The drawback of the nearest neighbor matchings is the complexity when matching two large pools of image features and the computative costly division for the NNDR matching.

List search approaches for matching of image features

The matching of two large pools of image features to find pixel correspondences in different images results in a costly process, because a correspondence measure and the first two nearest neighbors have to be evaluated for each possible feature combination. By restricting the pool of feature candidates for the matching process, a significant reduction of problem size is achievable. A possible restriction bases on feature properties, e.g., localization in the image, orientation or scale. Constraining the feature candidates means, that the pool of all image features has to be scanned for valid candidates, which is a list search problem.

1. **Sorted Linear Candidate Search**
   A prior sort of the pool regarding the restriction parameter enables a reduction in search time. By using the iterative successively approximation, the list index of the first element which fulfills the restriction is searched. The last candidate of the reduced list is searched with a linear search.
   After each iteration, the step size is halved and the search index is incremented or decremented depending on whether the restriction criterion is fullfilled. The initial step size is half the initial pool size.
2. **KD-Tree Candidate Search**
   A KD-tree [27] based search is a search tree with two edges per vertex and which divides the remaining set of feature candidates into two sets of the same size. By stepping through the KD-tree, the index of the first valid feature candidate is found efficiently. The disadvantage of this search method is the time consuming *a priori* construction of the KD-tree, which is not effective for small feature pools. In addition, if the

restriction search space has a low dimension, other search methods will perform faster.

8.2.3 Survey of Feature-based Self-Calibration

Extrinsic camera self-calibration is about recovering the extrinsic camera parameters using scene point correspondences only. Camera self-calibration is still a wide field of active research with different approaches. Early approaches are subdivided into aiming 3D reconstruction or not. The latter covers those algorithms where no information about the scene in front of the cameras is recovered during optimization.

One of the first approaches has been proposed by Longuet-Higgins [28]. The author introduced a linear method to recover the essential matrix, which is decomposable into the extrinsic parameters. Due to the required number of image point correspondences, it was introduced as the 8-point-algorithm.

Several following publications proposed optimizations regarding decomposition [29], plausibility [30, 31], and outlier handling for the corresponding image points [32]. As the linear approaches often lack the required accuracy, they are often followed by a non-linear refinement in a stratified process.

On the other hand, there are algorithms where camera parameters and 3D points of the scene are recovered simultaneously. One of those is bundle-adjustment [33]. Here a good initialization is required as Gauss-Newton optimization is involved. Thus, bundle adjustment is often chosen for the non-linear refinement as mentioned before.

Regarding online calibration procedures, they are classifiable as recursive or non-recursive. Recursive, or continuous self-calibration, means that temporal constraints are also optimized. Thus, image measurements in earlier time steps influence the current calibration result. Dang et al. proposed a parameter tracking system involving epipolar constraints and bundle adjustment [34]. In contrast to non-recursive self-calibration, there are no temporal constraints. Those are applied, in cases of a continuous decalibration or for active systems. Bjorkmann and Eklundh [35] introduced a real-time update of a restricted space of the extrinsic parameters. Pettersson and Petersson [36] extended a robust essential matrix estimation with a fast and robust FPGA-feature extraction. Parameter estimation for every new frame, beginning with rectified images, optimizing the extrinsic rotation and using a Kalman-Filter to limit overfitting was introduced by Hansen et al. [37].

8.3 Extraction of Image Features

Due to its stability and robustness, in respect of the requirements in advanced driver assistance systems, the Scale-Invariant Feature Transform (SIFT) by Lowe [9] is selected for this application as a state-of-the-art image feature descriptor and extractor in order to find sparse pixel correspondences in image pairs of a stereo camera system.

8.3.1 Detection of SIFT-Feature Points

Lowe's SIFT (Scale-Invariant Feature Transform, [9]) is a blob detector, which utilizes Lindeberg's scale-space approach [16] to achieve scale invariance. Blobs are detected by finding local maxima in the approximation of the Laplacian scale-space. The approximation of the Laplace operator is realized by the difference of two low pass filtered images, where both Gaussian kernels consist of different variances. The resulting scale-space approximation, the Difference of Gaussians (DoG), is constructed of several octaves with different image scales (see Figure 8.15). Every octave is subdivided into multiple intervals, which indicate the increasing variance of the Gaussian kernels. The initial interval of each octave arises by subsampling a specific interval of the previous octave. The DoG-pyramid, which represents the edges on multiples scales and different granularities, is browsed for local maxima in three dimensions (image position and intervals). After the detection of feature candidates in the discrete scale-space, their localization is refined by a Taylor series in order to position the candidates with subpixel accuracy and to approximate the extrema in the continuous scale-space.

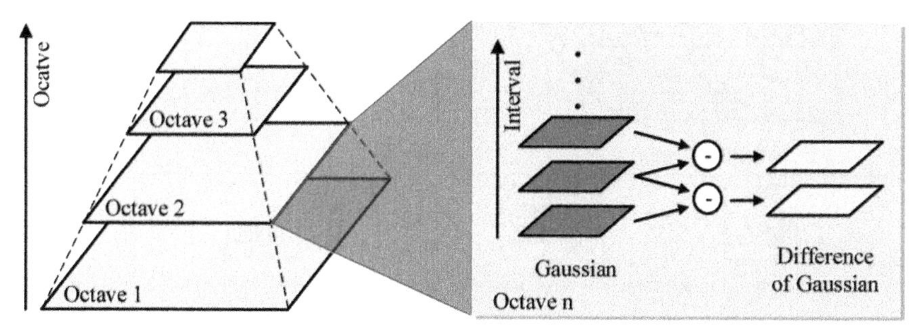

Figure 8.15 Image pyramid. The scale-space is constructed by different octaves, which consists of multiple intervals. Each interval indicates a specific variant of the used Gaussian kernel. In order to approximate the Laplace scale-space, the Difference of Gaussian is determined.

Candidates with a low contrast behavior and too edge like candidates are discarded.

8.3.2 Description of SIFT-Image Features

The SIFT-descriptor is a histogram-based descriptor and provides rotation invariance. Before histogram generation, the main orientation of each image feature is determined in order to align the local image region. To ascertain the main orientation for an image feature, a histogram of local image gradients is generated. The contribution of a local gradient to its corresponding orientation bin is defined by its magnitude and its distance to the feature point. After a smoothing step, the maximal histogram bin represents the main orientation of a feature point.

In addition to a reproducible detection of characteristic image points, a distinctive and robust description of the local neighborhood of the detected points is indispensable. For the description of image features, the gradient magnitude and orientation of the DoG-pyramid is used. A squared pixel area around the detected feature point is rotated by the feature orientation (see Figure 8.12) and subdivided into a grid (see Figure 8.16). For each

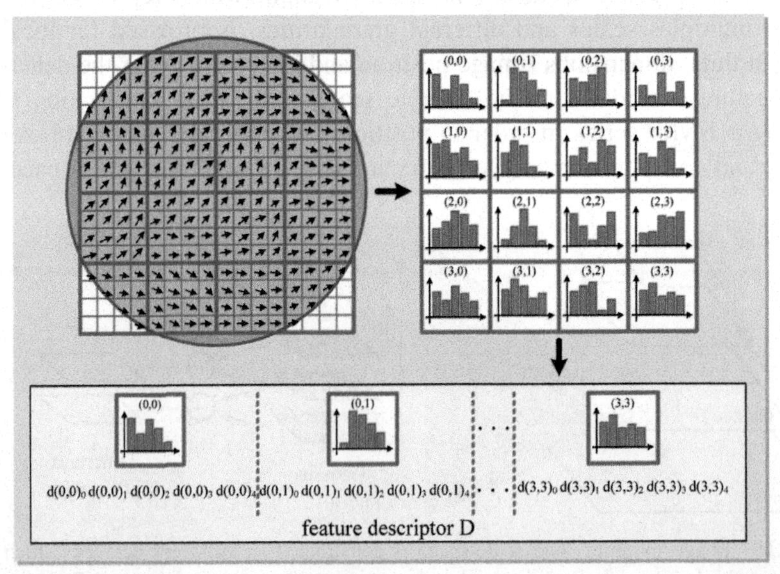

Figure 8.16 Generation of feature descriptor. The local neighborhood is subdivided into independent subregions, which are combined into individual histograms. After a weighting and smoothing, the feature descriptor is generated by concatenating the single histograms to as a resulting feature vector.

Figure 8.17 Extracted SIFT-features with exemplary geometry-based restriction of matching candidates. By restricting possible matching candidates geometrically, the problem size is significantly reduced.

grid element, an independent histogram of gradients is generated using orientation and magnitudes. The different histograms are weighted, smoothed and combined in a vector, which represents the final feature descriptor. The standard parameters of SIFT, which are suggested by Lowe [9], lead to 128 dimensions with floating point precision for the feature vector.

An exemplary SIFT-feature extraction of a rectified automotive scene is shown in Figure 8.17. The features of the left/right stereo camera are depicted in red/green. The scale of the features is illustrated as the circle's diameter, the orientation of the features with the additional radius line.

8.4 Matching of Image Features

The application of feature matching for advanced driver assistance systems favors correct pixel correspondences instead of a certain set of instable feature matches. Therefore, the matching of image features follows a straight forward approach with a significantly reduced problem size through matching of selected candidates. In this context, it is of minor interest which feature detector and extractor are used for the generation of image features.

Due to the fact, that SIFT is a histogram-based descriptor, a vector norm has to be evaluated as correspondence metric. A trade-off between computational complexity and conclusive results is the sum norm. The matching with sum norm results in marginally lower matching quality compared to matching with the Euclidean norm, but with a localization-based restriction of matching candidates, the matching results yield sufficient accuracy.

By constraining the pool of possible matching candidates, the problem size of feature matching is reduced significantly. The initial brute force matching requires a computation of the correspondence measure between each features of the left image and every feature of the right image. By taking into account the geometric set up of the stereo camera system, the search space is reduced to a fraction of the initial problem size, which results in a noticeable speed-up of matching and less wrong pixel correspondences at the same time (see Figure 8.17).

An exemplary result of the primarily brute force feature matching and for the enhanced matching process using the mentioned algorithmic setup is shown in Figure 8.18. Both stereo input images are overlaid and the image related features are displayed in red/green for the left/right stereo image. The significant increase of matching quality is expressed by the reduction of detected false pixel correspondences (blue connections) in relation to the correct pixel assignments (yellow connections). For the depicted results of feature matching, the sum norm is applied as correspondence measure and a localization-based restriction for choosing matching candidates is used.

Figure 8.18 Exemplary results of feature matching. The left and right stereo images are overlaid; features of the left/right image are displayed in red/green. Correct matches are depicted in yellow; false matches are shown in blue. The upper image shows the results of the initial brute force matching, whereas the lower image shows the results of the enhanced matching process.

8.5 Extrinsic Online Self-Calibration

Hartley and Zisserman present the fundamentals of extrinsic online self-calibration in their book [38] about multiple view geometry. The extrinsic parameters of a stereo system are described by the rotation $\mathrm{R_X} \in SO(3)$ and the translation vector $\mathrm{t_X} \in \mathbb{R}^3$. Given the extrinsic parameters the transformation of a point $\mathbf{X}_\mathrm{l} \in \mathbb{R}^3$ in the left camera coordinate system into the right camera coordinate system is described as

$$\mathbf{X}_\mathrm{r} = \mathrm{R_X}(\mathbf{X}_\mathrm{l} - \mathrm{t_X}).$$

Normally, extrinsic stereo camera calibration comes down to recovering $\mathrm{R_X}$ and $\mathrm{t_X}$. In the following, $\mathrm{t_X}$ is assumed constant and only $\mathrm{R_X}$ is recovered. During rectification $\mathrm{R_X}$ is broken down into

$$\mathrm{R_X} = \mathrm{R}_r^{-1} R_l$$

in order to determine the rotation of the left and right camera coordinate system to the common image plane respectively.

As decalibration is assumed to vary within a small range of only a few degrees, the recalibration is based on pre-rectified image point correspondences. The images may be pre-rectified using the camera parameters from the initial offline or a previous calibration run.

Given N as the corresponding pre-rectified image points $\widetilde{\mathrm{P}}_i$ and $\widetilde{\mathrm{Q}}_i$ for $i = 1, \ldots, N$ and assuming pinhole camera matrices K for simplicity, the image points are related to their unit directional image vectors

$$\tilde{p}_i \cong \mathrm{K}^{-1}\widetilde{\mathrm{P}}_i$$

$$\tilde{q}_i \cong \mathrm{K}^{-1}\widetilde{\mathrm{Q}}_i.$$

These vectors are related by the common epipolar constraint

$$0 = \widetilde{\mathrm{Q}}_i \mathrm{K}\,\check{\mathrm{R}}\,\mathrm{K}^{-1}\widetilde{\mathrm{P}}_i$$

whereas $\check{\mathrm{R}}$ denotes the rotation compensating the decalibration.

Since the decalibration is assumed to be small, optimization close to the identity matrix has to be avoided due to overfitting. Thus, the image vectors are re-rotated in the original camera coordinate systems via

$$\mathrm{p_i} = \mathrm{R_l}\widetilde{\mathrm{p}}_\mathrm{i}; \quad \mathrm{q_i} = \mathrm{R_r}\widetilde{\mathrm{q}}_\mathrm{i}.$$

Projecting them onto their respective image planes yields

$$\mathrm{P}_i = \mathrm{K}p_i; \quad \mathrm{Q}_i = \mathrm{K}q_i.$$

Given the measured image vector p_i, the depth d_i of the scene point $\mathbf{X}_\mathrm{l}$ and the decalibration $\check{\mathrm{R}}$, the corresponding image point Q_i may also be modelled as

$$\mathrm{Q}'_i(\check{\mathrm{R}}, \mathrm{d}_i) = \mathrm{K}\check{\mathrm{R}}\mathrm{R}_\mathrm{X}((p_i \mathrm{d}_i) - \mathrm{t}_\mathrm{X}).$$

Due to noise there is no exact solution, the objective function has to minimize the reprojection error e_i between measured and modelled image points

$$e_i = \|\mathrm{Q}_i - \mathrm{Q}'_i(\check{\mathrm{R}}, \mathrm{d}_i)\|.$$

Thus, the objective function including all image point correspondences is to minimize the sum of all squared reprojection errors and is formulated by

$$\underset{\check{\mathrm{R}},\,\mathbf{d}}{\operatorname{argmin}} \sum_{i=1}^{N} e_i^2$$

with $\mathbf{d} = [\mathrm{d}_1 \ldots \mathrm{d}_N]$. The solution is found by a non-linear optimization method, e.g., Levenberg-Marquardt.

8.6 Application-Specific Algorithmic Parameterization

The manifold varieties of algorithmic parameterizations for feature-based camera self-calibration lead to a sprawling design space, which is barely ascertainable in its entirety. Two exemplary selected application-specific aspects out of this design space are presented in this section. In subsection 8.6.1, the impact of differing bit depth of input images on the extraction of SIFT-features is shown. The parameterization of the presented matching methods is discussed in subsection 8.6.2.

8.6.1 Decreasing Bit Depth of Input Images for Extraction of SIFT-features

The availability of various cameras and the ongoing development of image processor technology lead to stereo systems, which provide digital images with a higher dynamic range. A higher bit depth of 8, 12 or 16 bit per pixel (bpp) promises a higher degree of representable details. However, it is not proven that a feature extractor will extract features of higher quality, when the bit depth for the input images is increased. In case of SIFT-feature extraction for a stereo camera self-calibration, this section shows, that the extracted pixel correspondences for 8 bpp input images and 12 bpp input images lead to identical pixel correspondences.

To ensure full accuracy during computations and to avoid effects of application-specific optimizations, a floating point software version of the SIFT-feature extraction is fed with 8 bpp and 12 bpp input images. Depending on the pixel depth of the input images, a bit depth specific algorithmic parameter set is configured.

After the SIFT-feature extraction, the nearest-neighbor distance ratio matching in combination with a geometry-based restriction of matching candidates (GB NNDR) is applied in order to find corresponding pixels. The experiment is accomplished with a dataset for which rectified input images and related disparity maps exist to validate the detected pixel combinations (see Figure 8.19). By checking the disparity of a match position in the left input image, it is possible to verify the corresponding match position in the right image. A radius offset for the detected matches of $\varepsilon = 0.5$ pixels for the position is tolerated during this investigation. The quantities for the extracted features and detected matches are shown in Table 8.3. The algorithmic parameters for the different SIFT-feature extractions are chosen to yield at least 1,000 features for both input images of the stereo camera system.

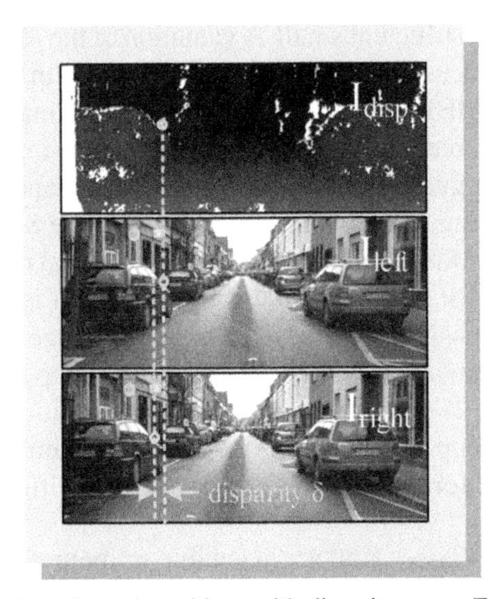

Figure 8.19 Verification of match positions with disparity maps. For rectified images, the horizontal difference of feature positions of a corresponding pixel pair equals the related value of the disparity map. With this technique, it is possible to validate resulting matching lists for datasets with ground truth disparity maps.

Table 8.3 Numbers of extracted SIFT-features and detected matches for 8 bpp input images and 12 bpp images. The number of the geometry-based (GB) nearest-neighbor distance ratio matches (NNDR) drops significantly but ensures a high explicitness of matches. The algorithmic parameters of the SIFT-feature extraction of the two test cases are adjusted in order to extract a similar number of features, which lead to an identical number of verified matches

	8 bpp Image	12 bpp Image
#SIFT-features left image	1,056	1,069
#SIFT-features right image	1,011	1,019
#GB NNB matches	1,013*	1,026*
#GB NNDR matches	608/60.0%	611/59.6%
#disparity verified matches	542/89.1%	544/89.0%
#matches not valid for evaluation	29/4.8%	28/4.9%
#matches wrong correspondences	37/6.1%	39/6.4%

**n* features of the left image have matched with features of the right image; duplicate assignments in the right image possible.

The significant difference between the number of geometry-based NNB matches and geometry-based NNDR matches is caused by the ratio factor, by which equivocal correspondences are rejected. A few correct pixel assignments may be rejected as well using this method, but the matching difference of those pixel pairs is not sufficient small. A valuation of the resulting absolute numbers is beyond the focus of this chapter, but by comparing the differences of the two versions of SIFT-feature extraction and matching it is clear, that there is nearly no difference between using an 8 bpp input image or a 12 bpp input image. To guarantee identical pixel correspondences, a visual inspection of the matching results is mandatory. In Figure 8.20 the result of detected SIFT-features of the left input image (blue: identical matches, orange: exclusive 12 bpp features, red: exclusive 8 bpp features) is shown. Out of 1,069 detected feature positions in the 12 bpp input image, 1,045 (97.8%) identical feature positions are detected again in the 8 bpp input image. In addition, there are 24 (2.2%) exclusive 12 bpp feature positions detected and 14 (1.3%) exclusive 8 bpp feature positions detected. Similar numbers are revealed by comparison for the feature extraction of the different right input images.

After the geometry-based NNDR matching of both feature sets, the comparison of the resulting pairs of the matched pixel correspondences allows a conclusion, if there is a difference between a feature extraction and matching of a 12 bpp input image and a 8 bpp input image. As shown in Figure 8.21, the bulk of the pixel correspondences are identical (blue lines); out of 611 found

Figure 8.20 Comparison of the resulting SIFT-features of the left input image for 12 bpp images and 8 bpp images. In the 12 bpp input image, an overall number of 1,069 features have been detected, whereas in the 8 bpp input image 1,056 features have been determined. A subset of 1,045 features (97.8%) is identical in both images (blue). There are 14 (1.3%) exclusive 8 bpp feature positions (red) detected and 24 (2.2%) exclusive 12 bpp feature positions (orange).

Figure 8.21 Comparison of the resulting pixel correspondences for the 8 bpp and 12 bpp input images. In the 12 bpp input image, an overall number of 611 pixel pairs has been detected, whereas in the 8 bpp input image 608 correspondences have been determined. A subset of 587 pairs (96.1%) is identical in both images (blue lines). Furthermore, there are 23 (3.8%) exclusive 8 bpp pairs (red lines) and 24 (3.9%) exclusive 12 bpp pixel correspondences (orange lines).

correspondences, 587 pairs (96.1%) are equal. In addition, there are 23 (3.8%) exclusive 8 bpp correspondences (red lines) and 24 (3.9%) exclusive 12 bpp correspondences (orange lines).

By tuning the algorithmic parameters in relation to the pixel depth of the used input images in this case study, it is possible to extract identical pixel correspondences. If there is no reason for further image processing steps, which require a proven higher bit depth than an 8 bpp graymap image, it is advisable to process the standard 8 bpp image in order to save computation resources.

8.6.2 Threshold-based Feature Matching

In this context of wide baseline stereo matching, threshold-based feature matching is used. As highlighted in subsection 8.2.2 , a nearest-neighbor-based match is defined as a pair of two descriptors, which are nearest neighbors of a matching process with a descriptor distance below a threshold. Furthermore, a feature only has one matching correspondence. In order to ensure a high rate of correct matches with a low rate of false matches, simultaneously, the threshold has to be selected in accordance to the algorithmic setup and the application-specific image content. Therefore, in this section a method for threshold selection is presented.

Underlying assumption for selecting a threshold for the presented NNB matching is the fact that there are correct matches with a low descriptor distance, false matches with a higher descriptor distance and nothing in between. Again, correct and false matches in this experiment are evaluated with existing disparity maps of the stereo camera system. The descriptor distances of an idealized NNB feature matching is shown in Figure 8.22 (right

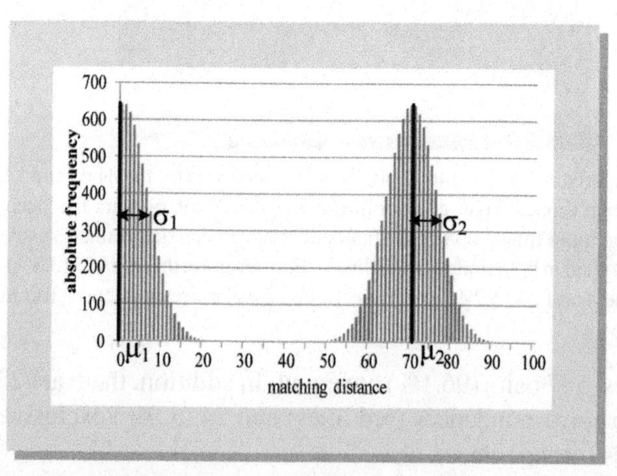

Figure 8.22 Histogram of random generated SIFT-descriptor distances of an idealized NNB feature matching. The right distribution with mean μ_2 displays the distances of wrong matches, whereas the left distribution with mean μ_1 illustrates the correct matches.

plot). For this experiment, 2×10^6 random generated SIFT-descriptors have been generated, pairs have been matched and the distances have been evaluated in a histogram. The resulting distribution of descriptor distances equals the Gaussian distribution, defined by mean μ_2 and deviation σ_2. Obviously, those descriptor distances are false matches. Correct matches follow the same distribution, but with differing mean μ_1 and deviation σ_1, as depicted in Figure 8.22 (left plot). By definition, descriptor distances are sums of absolute values, negative distances are not possible.

By comparing the distance histogram of the synthetic idealized NNB feature matching (see Figure 8.22) with a real-world NNB SIFT-feature matching (see Figure 8.23, left plot), two distinctive differences are noticeable: Firstly, the distance distribution for the correct feature distances and the false feature distance are overlapped and secondly, both distributions are skewed in direction of the others distribution mean value. This distortion is explainable by the fact, that there are always non-avoidable false positives and false negatives during the matching process. Further information concerning the distance distribution is available in [39].

The resulting distance distribution for the NNB SIFT-feature matching is shown in Figure 8.23 (right plot). Based on this plot, a suitable threshold for the matching process has to be extracted. It is desirable to select a threshold, which skips all of the false matches and approves all correct matches, and which corresponds to a threshold between the two ideal distributions. Due to skewing and overlapping of the distributions, there is always a set of false matches, which has to be tolerated by the chosen threshold. Therefore, the

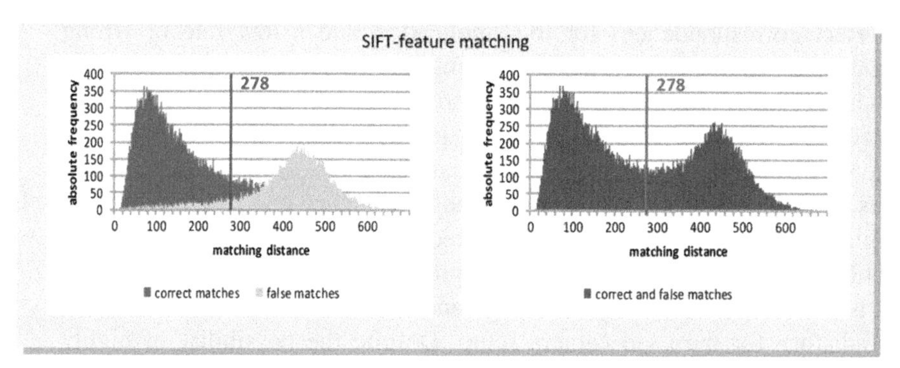

Figure 8.23 Histogram of descriptor distances for a NNB SIFT-feature matching with the extracted threshold according to Otsu. Distances of correct/wrong matches are displayed in blue/orange. The complete distribution is shown in purple.

goal is to minimize the false matches and maximize the correct matches, simultaneously.

Using the Otsu method [40], two overlapping distributions are separable by applying the discriminant criterion and utilizing the zeroth- and first-order cumulative moments of the distance histogram. Originally, Otsu presented his method for binarization of grey scale images, but the algorithm may be generalized for different types of histogram decomposition. By separating the two Gaussian distributions with Otsu's method, the descriptor distance which divides the distribution into a correct and a false region is determined and set as the matching threshold. Four different case studies have been executed (see Figures 8.23 and 8.24). Even for distance distributions, which do not show such a clear composition of two Gaussian distributions as the SIFT-feature matching case demonstrates, the Otsu's applied method provides reasonable thresholds.

For the entire application of wide baseline stereo matching, the threshold extraction has been performed offline, but it is also conceivable to implement an adaptive frame-to-frame online threshold extraction.

8.6.3 Parameterization of Matching Methods

The aim of this section is the evaluation of the presented matching procedures (see subsection 8.2.2) and the related parameter sets regarding their quality of assigned pixel correspondences in stereo camera systems images. The presented matching methods (TB, NNB, NNDR) result in varying correspondence lists, each of different size and with a variable percentage of correct pixel correspondences. The matching technique, which provides a high rate of correct correspondences for this application and a low rate of wrong assignments simultaneously, has to be identified.

It is possible to speed up the matching process through helpful assumptions about the position of corresponding feature points based on the given geometry of the stereo camera system. Using a spatial pre-selection of detected feature points, the number of candidates for the subsequent descriptor matching is significantly limited. In addition to reducing the problem size for the matching step, the quality of the feature point correspondences is increased. This is caused by excluding matching candidates, which are geometrically contradictory for the used camera setup. Despite the possibility of highly similar descriptors, wrong correspondences are prohibited even before the matching step using this technique.

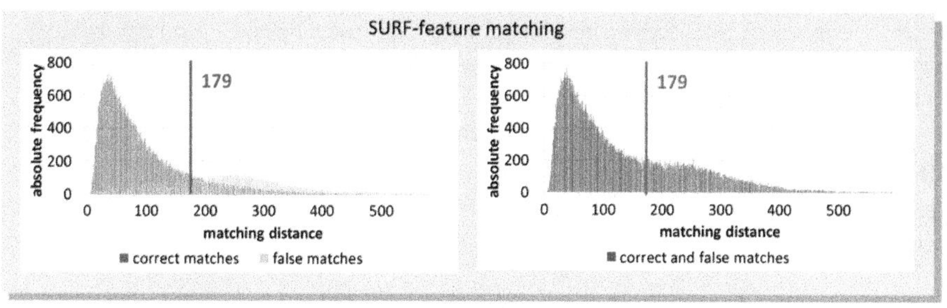

(a) Case study: SURF-feature matching

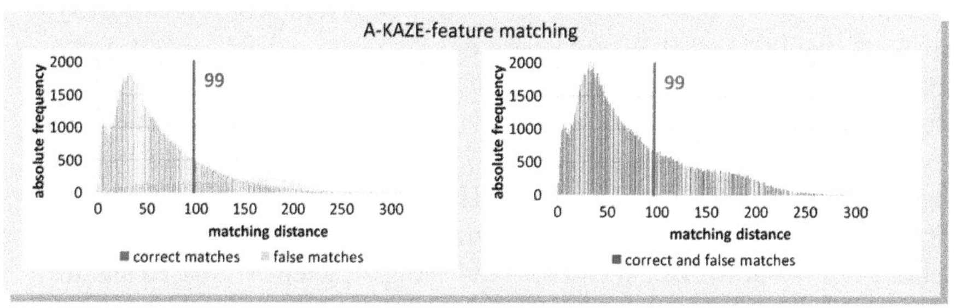

(b) Case study: A-KAZE-feature matching

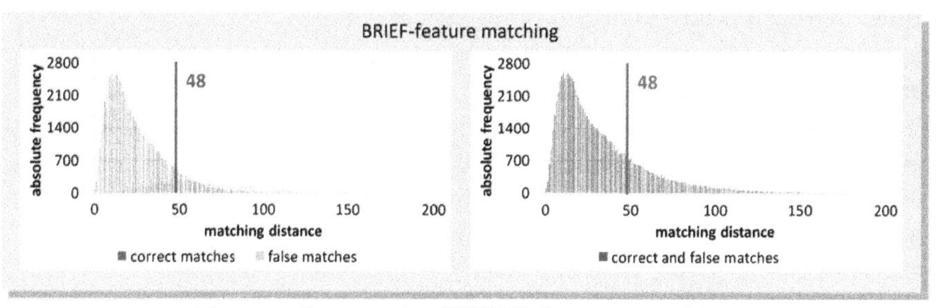

(c) Case study: BRIEF-feature matching

Figure 8.24 Histograms of descriptor distances for different NNB feature matching case studies with the extracted threshold according to Otsu. Distances of correct/wrong matches are displayed in blue/orange. The complete distribution is shown in purple. Due to different descriptors and resulting matching distances, various axis scales for clear presentation are used.

Geometry-based feature matching

The effect of spatial restriction of possible matching candidates (see Figure 8.17) in order to reduce the problem size for the feature matching depends on the permissible window size for matching candidates. In Table 8.4,

Table 8.4 Results for a SIFT-feature matching for a global matching and a geometry-based feature matching. The window size for the geometry-based feature matching is $+/-4$ pixel in y-direction and $+100/-4$ pixel in x-direction

	Global Matching	Geometry-Based Matching
#SIFT-features left image	1,057	1,057
#SIFT-features right image	1,011	1,011
#avg matching candidates	**1,011@1,057 matchings**	**7@1,057 matchings**

an overview of the average number of candidates per matching event is given. In the left/right 8-bit input image, 1,057/1,011 SIFT-features are extracted, which leads to $1{,}011\times1{,}057$ descriptor comparisons, when a brute force approach is used. With a window size of $+/-4$ pixel in y-direction (for rectified input images) and $+100/-4$ pixel in x-direction, the average number of descriptor comparisons is reduced to $7\times1{,}057$, which is a reduction of problem size of two orders of magnitude. The exact numbers of the candidate distribution for the geometry-based matching are shown in Figure 8.25. The reduction of problem size by a factor of $\times144$ using the geometry-based feature matching in relation to the global matching clearly outperforms the test, if a detected feature is a matching candidate. Therefore, using the geometry-based matching approach is advisable.

Choosing a matching method

Different methods of feature matching with or without a spatial restriction of the matching candidates directly affect the quality of resulting feature correspondence lists. Exemplary numbers for a variation of matching methods are shown in Table 8.5. Again, in the left/right 8-bit input image, 1,057/1,011

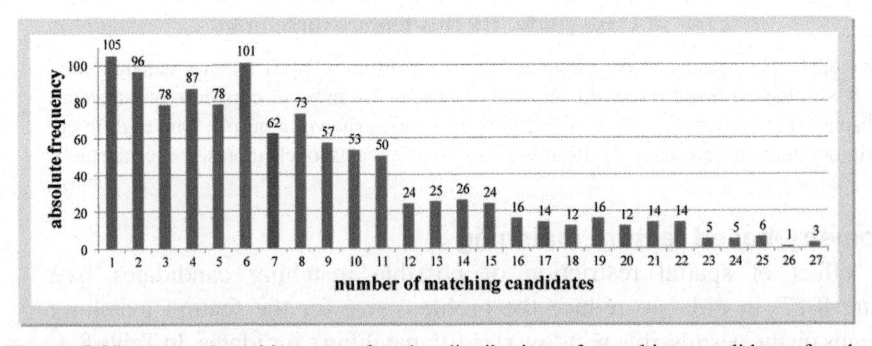

Figure 8.25 Exemplary histogram for the distribution of matching candidates for the geometry-based feature matching (see Table 8.4). The average number of candidates is 7 candidates per matching event.

Table 8.5 Results of disparity verified feature correspondences for different combinations of global and spatial restriction matching methods. In addition to a high rate of correct matches, a minimal number of pixel correspondences has to be given for a reliable subsequent image processing. The total numbers of detected matches for selected algorithmic combinations are given in brackets. The number of correct matches and wrong matches do not result in 100% because of missing values in the ground truth disparity maps. Those values are skipped for evaluation

	Global Matching		Geometry-Based Matching	
	#Correct Matches	#Wrong Matches	#Correct Matches	#Wrong Matches
#TB disparity verified matches	562 (1,057)	400	702 (1,006)	240
	53.2%	37.8%	69.8%	23.9%
#NNB disparity verified matches	541 (735)	149	**556 (597)**	**22**
	73.6%	20.3%	**93.1%**	**3.7%**
#NNDR disparity	493 (540)	31	542 (605)	36
verified matches	91.3%	5.7%	89.6%	6.0%

SIFT-features are extracted. The total number of detected matches for each algorithmic combination is given in brackets (see Table 8.5).

For each method, the geometry-based feature matching grants an improvement of the correct matching rate or the rate remains in the same order of magnitude. The resulting correspondence lists generated with the threshold-based feature matching has the highest number of entries, but the quota of correct matches is insufficiently low. A combination of NNB-matching and the geometry-based restriction leads to the highest rate of correct matches (93.1%) and a low rate of wrong matches (3.7%), simultaneously. Furthermore, the absolute number of correct matches (556) guarantees a stable base for following image processing algorithms. Therefore, the use of the NNB-matching with a geometry-based restriction of matching candidates in order to extract pixel correspondence lists for a feature-based camera self-calibration is recommended.

Accuracy of localization

All prior investigations in this section are based on the assumption that 'disparity verified matching' defines the consensus of the extracted feature-based disparity including a small offset ε and the related actual disparity taken from the disparity ground truth map. This offset ε is necessary in order to tolerate small deviations of feature positions, which are caused during the localization step.

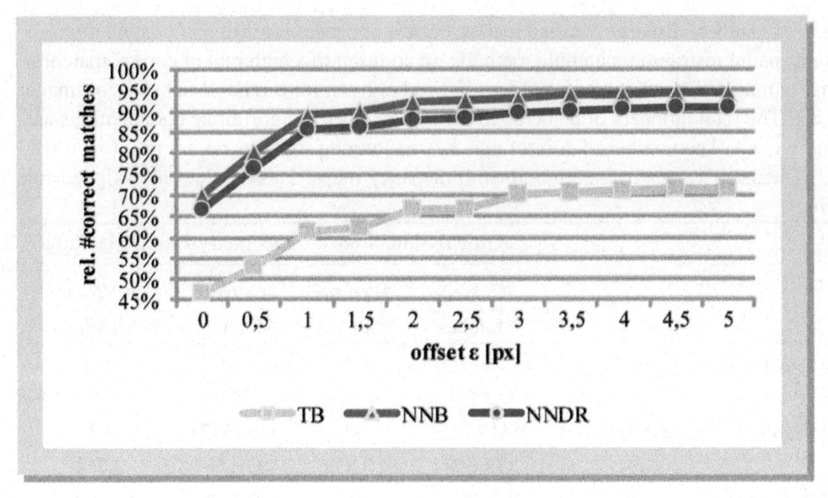

Figure 8.26 Rates of disparity verified pixel correspondences for different offsets ε and three matching methods. For all methods, the rate of correct matches runs into saturation. The NNB matching method performs best over all offsets ε. (TB: Threshold-Based Matching; NNB: Nearest-Neighbor-Based Matching; NNDR: Nearest-Neighbor Distance Ratio Matching).

To evaluate the impact of varying offsets ε, in the left/right 8-bit input image, 1,057/1,011 SIFT-features are extracted and matched with a geometry-based approach for the TB, NNB and NNDR matching method. The rates of disparity verified pixel correspondences for different offsets ε are shown in Figure 8.26. Remarkably the qualitative trend is identical for all matching methods. Furthermore, all methods run into saturation for offsets higher than 3 pixel. As expected, the threshold-based matching (TB) provides the lowest matching rate for all offsets ε. The nearest-neighbor based (NNB) matching method results constantly in the highest rate for disparity verified matches with approximately over 90% (<537 out of 597 matches) for an offset larger than 1 pixel. It is worth mentioning that 70% of all matches (419 out of 597 matches) for the NNB method are identical to the ground truth disparity map (offsets $\varepsilon = 0$ pixel).

To achieve an applicable trade-off between exact 'disparity verified correspondences' and permitting localization errors due to viewpoint changes, all prior investigations have been verified with an offset $\varepsilon = 3$ pixel.

8.7 Hardware Based SIFT-Feature Extraction

Fast and reliable extraction of SIFT-features in the presented context of feature-based camera self-calibration requires a tuned implementation of the algorithm for the hardware platform used. Therefore, in this section, the relevant hardware properties of SIFT-feature extraction are introduced and an overview of existing SIFT-feature implementations is given.

8.7.1 Challenges of SIFT-Feature Extraction

The extraction of SIFT-features is a challenging task due to the number of operations and memory accesses that have to be executed. As depicted in Figure 8.27, the algorithmic steps of SIFT-feature extraction differ in varying ratios of control complexity and regular arithmetic. As shown in [41], the building of scale-space, which consists of multiple separable and symmetric Gaussian filters, is an arithmetically intensive task with almost no control overhead. In contrast, parts of the feature points detection or the descriptor generation require control mechanisms, which result in heavy branching on conventional processors. Furthermore, the scale-space is mandatory for the feature description and has to be buffered until the generation of descriptors, which requires a large memory and arbitrarily non-aligned memory accesses aggravate the challenging memory bottleneck. In addition, the algorithmic quality of SIFT has to be ensured for subsequent processing steps, which requires an appropriate level of internal accuracy of the temporal results.

Therefore, specialized architectures are necessary to ensure the processing performance demanded for SIFT-feature extraction. At the same time, those specialized systems have to be as flexible as possible to guarantee a fast

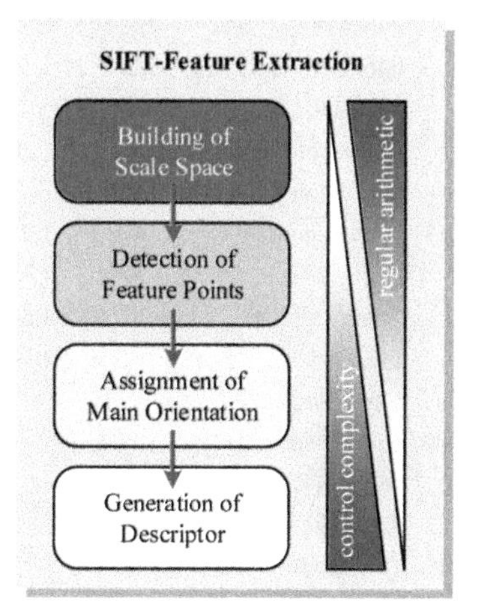

Figure 8.27 Break down of SIFT-feature extraction into four algorithmic steps and relating qualitatively quota of control complexity and complexity (i.e., regular arithmetic).

implementation of future algorithms which might perform better compared to state-of-the-art feature extractors [42].

8.7.2 Existing Systems for Hardware Based SIFT-Feature Extraction

In the following Table 8.6, a set of existing systems/platforms for the hardware based SIFT-feature extraction is presented. The selection shown is not meant to be exhaustive, but elucidates the trade-off of different platforms regarding sufficient processing power, low power consumption and satisfactory flexibility for future algorithm implementations.

Moren et al. [43] presented in 2015 a comprehensive survey of a SIFT-feature extraction for homogeneous and heterogeneous CPU/GPU systems. With different techniques for parallelization and a portable performance concept using OpenCL (Open Computing Language), the SIFT-feature extraction has been implemented on various single device and multi-device platforms.

Table 8.6 Overview of existing systems for SIFT-feature extraction

	Implementation				Frequency	Performance
	Author	Year	Device	Powre	(MHz)**	(fps)
CPU & GPU	Moren et al. [43]	2015	Nvidia GTX 780 TI	250 W*	875	137.6 @ 640x480
			AMD R9-290	300 W*	947	98.7 @ 640x480
			Nvidia GTX 580	244 W*	772	77.2 @ 640x480
			Nvidia Tesla C2050	238 W*	1150	74.0 @ 640x480
			Intel MIC 3120A	300 W*	1100	16.8 @ 640x480
			Intel Core-i7 4930K	130 W*	3400	32.6 @ 640x480
			Intel Xeon E5-2667	130 W*	2900	28.3 @ 640x480
			AMD Opteron 6168	115 W*	1900	8.0 @ 640x480
			Intel Xeon E5-2667	130 W*	2900	4.0 @ 640x480
Mobile GPU	Rister et al. [44]	2013	Snapdragon S4	~4 W	1,700/400	9.9 @ 320x240
			Nexus 7	N/A	1,600/520	8.6 @ 320x240
			Galaxy Note II	N/A	1,600/400	7.6 @ 320x240
			Tegra 250	~3 W	1,000/333	7.9 @ 320x240
FPGA & ASIC	Bonato et al. [45]	2008	Altera Stratix II	N/A	100	30.0 @ 320x240
	Yao et al. [46]	2009	Xilinx Virtex 5	N/A	100	32.3 @ 640x480
	Huang et al. [47]	2012	TSMC 18μm CMOS	N/A	100	30.0 @ 640x480
	Yum et al. [48]	2015	Xilinx Virtex 6	N/A	170	36.9 @ 1280x720
ASIP	Mentzer et al. [41, 42]	2015	TSMC 45nm process	<1 W	400	1 @ 800x640

*Thermal Design Power.

**For category mobile GPU: CPU/GPU frequency.

The systems are separated into four different implementations, where each implementation is optimized according to device specific characteristics:

- Host-device implementation for control
- GPU device implementation
- Multi-core CPU device
- Multi-device implementation

The systems are evaluated for multiple image sizes for equal algorithmic setups. Single device runtimes are listed in Table 8.6 for VGA image size. Noticeable is the fact, that all single GPU systems and multi-device systems, in which a GPU is enlisted, provide enough performance for a real-time SIFT-feature extraction for VGA images, but require more than 230 W power consumption. Furthermore, CPU single device systems are close to real-time by providing 17–32 fps, but again, the power consumption is far too high for use in automobiles with over 115 W power consumption. The AMD Opteron 6168 and Intel Xeon E5 do not reach a sufficient frame rate for a SIFT-feature extraction application. The author presents three different heterogeneous systems, which are assembled by the afore mentioned single device systems, which provide enough performance for real-time applications even for very large images. For all systems, the flexibility is ensured by using the high-level OpenCL.

In 2013, Rister [44] proposed an investigation of SIFT-feature extraction on four different platforms using mobile GPUs. The author used a heterogeneous dataflow scheme and applied a partitioning of workload between CPU and GPU. Different platform specific optimizations are used, e.g., data compressing by pixel reordering or branchless convolution through on-the-fly code generation. With frame rates reaching between 7.6 fps and 9.9 fps, the performance is too poor for an use in ADAS, but a power consumption of the complete systems of <5 W fulfills the requirements demanded. Furthermore, flexibility is guaranteed by OpenGL for Android.

Bonato et al. presented 2008 the first hardware based SIFT implementation [45]. The heterogeneous system consists of a hardware accelerator for SIFT-feature detection and a NIOS II softcore processor for SIFT-descriptor generation. The system has been emulated on an Altera Stratix II FPGA and a frame rate of 30 fps for QVGA images has been reached.

One year later in 2009, Yao et al. claimed to reach a comparable frame rate of 32.3 fps, but for VGA images. They presented a hardware-based SIFT-feature detector, which has been emulated on a ML507 board, and a SIFT-feature generation in software. The drawback of the presented work is the simplified SIFT scale-space, which leads to a limited algorithmic quality, compared to the original algorithm.

The first fully hardware-based SIFT-feature extraction has been presented in 2012 by Huang et al. [47]. The author's system reaches a frame rate of 30 fps for VGA images and uses a TSMC 180 μm CMOS process.

In 2015, Yum et al. proposed a FPGA-based full SIFT implementation, which is capable of processing 36.85 fps for HD images on a Xilinx Virtex 6 device [48]. By reducing the amount of necessary internal memory and a local-patch reuse scheme, a high data throughput is reached, but the building of scale-space is adjusted, which affects the algorithmic quality.

These hardware-based approaches provide adequate processing power for a high frame rate and a sufficiently low power consumption of typically <10 W, but the presented systems are not SW-flexible.

Mentzer et al. [41, 42] presented an ASIP-based SIFT-feature extraction, which preserves the full algorithmic quality. Sufficient flexibility for future algorithms of image feature extraction is ensured by the platform-specific attribute of full software programmability. The drawback of the presented case study is the low frame rate in FPGA emulation, which prohibits a real time application in automotive use.

Thus, heterogeneous systems consisting of dedicated hardware for accelerating the scale-space construction and a processor-based descriptor generation is a promising trade-off between flexibility, performance and power consumption. State-of-the-art conventional CPUs and GPUs are too power greedy, nowadays mobile GPUs do not reach sufficient frame rates and pure hardware-based systems do not fulfill the requirements for flexibility. A trade-off concerning flexibility by supporting a processor with non-programmable hardware accelerators is a possible approach for a SIFT-feature extraction in the field of Advanced Driver Assistance Systems.

8.8 Conclusion

In this chapter, selected aspects of self-calibration for wide baseline stereo camera systems for automotive applications have been introduced. Starting at the extraction and matching of image features up to the extrinsic online self calibration of stereo camera systems, fundamental algorithms have been presented. A promising algorithmic combination consisting of the extraction of SIFT-features, nearest-neighbor-based matching with spatial selection of matching candidates and the estimation of camera parameters in order to rectify misaligned stereo images have been discussed in detail.

Three exemplary aspects of algorithmic parameterizations, which are the impact of a decreasing bit depth of input images, the selection of a matching method and the threshold selection for the matching process, have been examined in detail to show substitutionally the complexity of adjusting existing algorithms to new applications.

In the last section, basic challenges of hardware-based SIFT-feature extraction are presented and hardware-specific solutions for the afore mentioned algorithmic challenges are discussed. Finally, existing systems for the extraction of SIFT-features are reviewed.

As discussed in this chapter, there is no state-of-the-art hardware implementation for the proposed algorithmic combination, which fulfills the three requirements for ADAS, and delivers sufficient processing performance, low power consumption and full flexibility for future algorithms. Thus, remaining challenges will be solved to improve safety for vulnerable road users and to enhance comfort in future automobiles.

References

[1] K. Genuit, Sound-Engineering im Automobilbereich, Springer, 2010.

[2] C. Schmid, R. Mohr and C. Bauckhage, "Evaluation of Interest Point Detectors," *International Journal of Computer Vision*, pp. 151–172, 2010.

[3] K. Mikolajczyk and C. Schmid, "A Performance Evaluation of Local Descriptors," *IEEE Transcations on Pattern Analysis and Machine Intelligence,* pp. 1615–1630, 2005.

[4] H. Aanæs, A. L. Dahl and K. S. Pedersen, "Interesting Interest Points," *International Journal of Computer Vision,* pp. 18–35, 2012.

[5] S. Gauglitz, T. Höllerer and M. Turk, "Evaluation of Interest Point Detectors and Feature Descriptors for Visual Tracking," *International Journal of Computer Vision,* 2011.

[6] J. Heinly, E. Dunn and J.-M. Frahm, "Comparative Evaluation of Binary Features," *ECCV,* pp. 759–773, 2012.

[7] K. Mikolajczyk and C. Schmid, "Scale & Affine Invariant Interest Point Detectors," *International Journal of Computer Vision,* pp. 63–86, 2004.

[8] J. Shi and C. Tomasi, "Good Features to Track," *Proceedings of the IEEE Conference on Computer Vision and Pattern Recognition (CVPR),* pp. 593–600, 1994.

[9] D. G. Lowe, "Object recognition from local scale-invariant features," *Proceedings of the International Conference on Computer Vision,* pp. 1150–1157, 1999.

[10] H. Hirschmüller, "Accurate and efficient stereo processing by semi-global matching and mutual information," in *IEEE Computer Society Conference on Computer Vision and Pattern Recognition*, 2005.
[11] Z. Zhang, "A Flexible New Technique for Camera Calibration," *IEEE Transactions on Pattern Analysis and Machine Intelligence*, pp. 1330–1334, 2000.
[12] K. Mikolajczyk, T. Tuytelaars, C. Schmid, A. Zisserman, J. Matas, F. Schaffalitzky, T. Kadir and L. V. Gool, "A Comparison of Affine Region Detectors," *International Journal of Computer Vision,* pp. 43–72, 2005.
[13] J. Canny, "A Computational Approach to Edge Detection," *Transactions on Pattern Analysis and Machine Intelligence,* 1986.
[14] C. Harris and M. Stephens, "A Combined Corner and Edge Detector," *Proceedings of the Alvey Vision Conference,* pp. 23.1–23.6, 1988.
[15] E. Rosten and T. Drummond, "Machine Learning for high-speed corner detection," *European Conference on Computer Vision,* 2006.
[16] T. Lindeberg, *Scale-Space Theory in Computer Vision*, Kluwer Academic Publishers, 1994.
[17] Y. Ke and R. Sukthankar, "PCA-SIFT: a more distinctive representation for local image descriptors," *Proceedings of Conference on Computer Vision and Pattern Recognition,* 2004.
[18] S. Belongie, J. Malik and J. Puzicha, "Shape Matching and Object Recognition Using Shape Contexts," *Transactions on Pattern Analysis and Machin Intelligence,* pp. 509–522, April 2002.
[19] E. Tola, V. Lepetit and P. Fua, "DAISY: An Efficient Dense Descriptor Applied to Wide-Baseline Stereo," *Transactions on Pattern Analysis and Machine Intelligence,* pp. 815–830, May 2010.
[20] M. Calonder, V. Lepetit, M. Ozuysal, T. Trzcinski, C. Strecha and P. Fua, "BRIEF: Computing a Local Binary Descriptor Very Fast," *IEEE Transactions on Pattern Analysis and Machine Intelligence,* pp. 1281–1298, 2012.
[21] A. Alahi, R. Ortiz and P. Vandergheynst, "FREAK: Fast Retina Keypoint," *Conference on Computer Vision and Pattern Recognition,* pp. 510–517, 2012.
[22] E. Rublee, V. Rabaud, K. Konolige and G. Bradski, "ORB: an efficient alternative to SIFT or SURF," *International Conference on Computer Vision,* pp. 2564–2571, 2011.
[23] S. Leutenegger, M. Chli and R. Y. Siegwart, "BRISK: Binary Robust Invariant Scalable Keypoints," *International Conference on Computer Vision,* pp. 2548–2555, 2011.

[24] P. F. Alcantarilla, A. Bartoli and A. J. Davison, "KAZE Features," *Proceedings of the 12th European Conference on Computer Vision,* pp. 214–227, 2012.

[25] H. Bay, A. Ess, T. Tuytelaars and L. Van Gool, "SURF: Speeded Up Robust Features," *Journal of Computer Vision and Image Understanding,* pp. 346–359, 6 2008.

[26] P. Alcantarilla, J. Nuevo and A. Bartoli, "Fast Explicit Diffusion for Accelerated Features in Nonlinear Scale Spaces," *Proceedings of the British Machine Vision Conference,* 2013.

[27] M. Muja and D. G. Lowe, "Fast Approximate Nearest Neighbors with Automatic Algorithm Configuration," *International Conference on Computer Vision Theory and Applications,* pp. 331–340, 2009.

[28] H. C. Longuet-Higgins, "A Computer Algorithm for Reconstructing a Scene from Two Projections," *Readings in Computer Vision: Issues, Problems, Principles, and Paradigms,* pp. 61–62, 1987.

[29] R. I. Hartley, "Estimation of relative camera positions for uncalibrated cameras," *European Conference on Computer Vision,* pp. 579–587, 1992.

[30] Z. Zhang, Q.-T. Luong and O. Faugeras, "Motion of an Uncalibrated Stereo Rig: Self-calibration and Metric Reconstruction," *IEEE Transactions on Robotics and Automation,* pp. 103–113, 1996.

[31] Q.-T. Luong and O. D. Faugeras, "Self-Calibration of a Moving Camera from Point Correspondences and Fundamental Matrices," *International Journal for Computer Vision,* pp. 261–289, 1997.

[32] P. Torr and A. Zisserman, "Robust Computation and Parametrization of Multiple View Relations," in *Computer Vision and Image Understanding*, 2000.

[33] B. Triggs, P. F. McLauchlan, R. I. Hartley and A. W. Fitzgibbon, "Bundle Adjustment – A Modern Synthesis," *Proceedings of the International Workshop on Vision Algorithms: Theory and Practice,* pp. 298–372, 2000.

[34] T. Dang, C. Hoffmann and C. Stiller, "Continuous Stereo Self-calibration by Camera Parameter Tracking," *Transactions on Image Processing*, pp. 1536–1550, 2009.

[35] M. Björkman and J.-O. Eklundh, "Real-Time Epipolar Geometry Estimation of Binocular Stereo Heads," *Transactions on Pattern Analysis and Machine Intelligence,* pp. 425–432, 2002.

[36] Petterson and Petterson, "Online stereo calibration using FPGAs," *Intelligent Vehicles Symposium,* 2005.

[37] P. Hansen, H. S. Alismail, P. Rander and B. Browning, "Online Continuous Stereo Extrinsic Parameter Estimation," *International Conference on Computer Vision and Pattern Recognition,* 2012.

[38] R. I. Hartley and A. Zisserman, Multiple View Geometry in Computer Vision, Cambridge University Press, 2004.

[39] R. Szeliski, Computer Vision: Algorithms and Applications, London: Springer-Verlag, 2011.

[40] N. Otsu, "A Threshold Selection Method from Gray-Level Histograms," *IEEE Transactions on Systems, Man and Cybernetics,* pp. 62–66, 1979.

[41] N. Mentzer, G. P. Vaya and H. Blume, "Analyzing the Performance-Hardware Trade-off of an ASIP-based SIFT Feature Extraction," *Journal of Signal Processing Systems,* 2015.

[42] N. Mentzer, N. V. Egloffstein, G. P. Vaya, W. Ritter and H. Blume, "Instruction-Set Extension for an ASIP-based SIFT Feature Extraction," in *International Conference on Embedded Computer Systems: Architectures, Modeling and Simulation (SAMOS)*, Samos, Greece, 2014.

[43] K. Moren and D. Göhringer, "A framework for accelerating local feature extraction with OpenCL on multi-core CPUs and co-processors," *Journal of Real-Time Image Processing,* pp. 1–18, 03 2016.

[44] B. Rister, G. Wang, M. Wu and J. R. Cavallaro, "A Fast and Efficient SIFT Detector using the mobile GPU," *IEEE International Conference on Acoustics, Speech and Signal Processing,* 2013.

[45] V. Bonato, E. Marques and G. A. Constantinides, "A Parallel Hardware Architecture for Scale and Rotation Invariant Feature Detection," *IEEE Transactions on Circuits and Systems for Video Technology*, pp. 1703–1712, 2008.

[46] L. Yao, H. Feng, Y. Zhu, Z. Jiang, D. Zhao and W. Feng, "An architecture of optimised SIFT feature detection for an FPGA implementation of an image matcher," *International Conference on Field-Programmable Technology,* pp. 30–37, 2009.

[47] F.-C. Huang, S. Huang, J. Ker and Y. Chen, "High-Performance SIFT Hardware Accelerator for Real-Time Image Feature Extraction," *IEEE Transactions on Circuits and Systems for Video Technology,* pp. 340–351, 03 2012.

[48] J. Yum, C.-H. Lee, J.-S. Kim and H.-J. Lee, "A Novel Hardware Architecture with Reduced Internal Memory for Real-time Extraction of SIFT in an HD Video," *IEEE Transactions on Circuits and Systems for Video Technology,* 2015.

9

Arbitration and Sharing Control Strategies in the Driving Process

David González[1], Joshué Pérez[1], Vicente Milanés[1], Fawzi Nashashibi[1], Marga Sáez Tort[2] and Angel Cuevas[2]

[1]INRIA, France
[2]CTAG – Centro Tecnológico de Automoción de Galicia, Spain

9.1 Introduction

Automated functions for real world traffic scenarios have been increasing in last years in the automotive industry. Many research contributions have been done in this field. However, other problems have come to the drivers, related to the legal and liability framework, where it is still unclear up to which point the control of the vehicle should stay with the driver or be taken by automation.

The aim of the Advanced Driver Assistance Systems (ADAS) is mainly related to help drivers in safety critical situations rather than to replace them. However, in recent years, many research advances have been done in this field, making automated driving closer to reality day by day. The numbers of automated driving functions for typical traffic scenarios have increased in the last few years in the automotive industry and university research. However, other problems have appeared for drivers of such automated cars: When should the driver or the automated systems take control of the vehicle (since both cannot control an automated vehicle together at the same time due to potential conflicts)? This question has not a simple answer; it depends on different conditions, such as: the environment, driver condition, vehicle capabilities, fault tolerance, among others. Arbitration and control activities have been implemented in DESERVE WP24, mainly motivated by this question.

In this chapter, we will analyze the acceptability to the ADAS functions available in the market, and its relation with the different control actions. A survey on arbitration and control solutions in ADAS is presented. It will allow to create the basis for future development of a generic ADAS control (the lateral and longitudinal behavior), based on the integration of the application request, the driver behavior and driving conditions in the framework of the DESERVE project. Based on vehicle modeling, driver behavior and intention, a first approach for arbitration and control strategies, which can anticipate the priorities on the control in emergency situations, is described.

The main aim of this work is to allow the development of a new generation of ADAS solutions where the control could be effectively shared between the vehicle and the driver. Some simulations will allow the virtual testing for the future implementation in demonstrators.

Fuzzy logic techniques are a suitable approach for the arbitration control in the driving process. The contributions described in this chapter will be implemented in two demonstrators: Automatic/Autonomous Emergency Braking (AEB) pedestrian protection system and Driver Distraction monitoring—CRF demo vehicles—using RTMaps[1] as the development software.

The proposed arbitration and shared control takes into account the state of the driver and the state of the system, in order to assess the level of control that each system should have; based on the standard SAE J3016. Fuzzy Logic controllers consider a control level that allows a smooth control sharing between the automated system and the driver. It has been design according to the Application Platform in DESERVE control architecture. Although the Fuzzy Logic (as some other Artificial Intelligence techniques) is not explicitly considered in the road vehicles functions safety standard (ISO 26262), a large number of applications have been developed in recent years. The behavior of a human driver can be emulated with this technique.

9.2 ADAS Functions Available in the Market

Driver Assistance Systems (DAS) or Advanced Driver Assistance Systems (ADAS) can be defined as those active safety systems which require some monitoring on the vehicle's environment and on driver intentions. This extra information is combined with ego-vehicle data (positions and speed profile) in order to provide the driver with some warning or perform some automatic

---

[1] https://intempora.com/

actuation with the goal of increasing safety. Regarding driver interactions, a DAS can offer:

- Information about the current situation
- A warning to alert the driver
- Take the control of the vehicle, partially or completely
- A combination of them

This section is focused on those DAS which have the capability of taking vehicle control to improve or correct the driver response.

From the control point of view, control DAS systems can be classified as:

- ***Longitudinal Control Systems***: Those DAS which are able to modify vehicle speed by accelerating or braking.
- ***Lateral Control Systems***: Those DAS which are able to change vehicle direction, usually actuating on the steering system.
- ***Global Control Systems***: DAS with a combination of longitudinal and lateral control.

The Control DAS examples described in this subchapter are shown below:

Longitudinal Control Systems

- ACC (Adaptive Cruise Control)
- FCW (Frontal Collision Warning or Forward Collision Warning)
- AEB/CMbB (Automatic Emergency Braking/Collision Mitigation by Braking)
- SLA (Speed Limit Assistant)

Lateral Control Systems

- LDW/LKA (Lane Departure Warning/Lane Keeping Assistance)
- BSD/LCA (Blind Spot Detection/Lane Change Assistant)

Other Control Systems

- Pedestrian Detection/Active Hood
- Driver Distraction Detection
- PreCrash
- Parking Assistance

9.2.1 Longitudinal Control Systems

These are the main steps for the longitudinal control of the vehicle: the first system is more a comfort than a safety one (ACC), but safety systems such as

Forward Collision Warning (FCW) or AEB are built upon it. Other possibilities for Longitudinal Control of the vehicle are systems such as SLA.

ACC (Adaptive Cruise Control)

The ACC adds to the most common Cruise Control constant safety distance maintenance with the preceding vehicle. It consists of a front-mounted sensor, an integrated control unit with the task to regulate the system's performance and a suitable HMI that informs and allows the driver to control the system.

This sensor controls the area in front of the vehicle. If no obstacle is detected, the vehicle keeps the selected speed as a standard cruise control. In case a vehicle is detected in the predicted path of the vehicle (target vehicle), the sensor calculates the relative distance and speed to the target vehicle. (up to around 150–200 m). Then, the Control Unit decides whether it is necessary to actuate the brake system of the vehicle with the goal to keep a constant safety distance. When the target vehicle disappears from the detection area, the Control Unit sends the order to accelerate again until the desired cruise speed is reached.

The system works usually between 30 and 180 km/h. The maximum deceleration provided by the system is far from the maximum deceleration capabilities of the vehicle (in between 2 and 3 m/s^2)[2]. The driver can choose between different safety gaps (time – related). Developed for high capacity

Figure 9.1 ACC Systems.

[2]In case the driver does not react, some other ACC systems are also improved with an AEB system, also considered as CMbB, providing autonomous brake action (from 5 m/s^2 to full power).

roads, ACC Stop & Go improves the performance of the conventional ACC to a full stop capability. The stop and go of the vehicle is, thus, automatically performed, so the range of the system is extended to 0–200 km/h.

FCW (Frontal Collision Warning)

When ACC fails to provide enough deceleration [exceed comfort specifications (above 2–3 m/s^2)], request to avoid a possible head-on collision, a warning, is provided to the driver (FCW). This warning reminds the driver the urge to take control of the situation. FCW is included in the basic ACC system in all vehicles equipped with the necessary sensors (laser, radar, etc.). These systems are usually activated between 5 and 2 seconds before the collision with the vehicle ahead might occur.

AEB/CMbB (Automatic Emergency Braking/Collision Mitigation by Braking)

As the third step in the longitudinal control of the vehicle, AEB is an automatic emergency safety system that takes control of the situation if the driver fails to decelerate the vehicle when a head-on collision is about to happen. The system consists on an automatic actuation of the vehicle's brakes in case the situation requires so to avoid a crash. AEB systems can be divided according to their deceleration in 1) *Soft Braking*. Up to 5 m/s^2 and 2) *Hard Braking*. From 5 m/s^2 to the full capability of the braking system.

Some systems can provide a progressive braking: first, a *soft braking* can be provided and, in case the accident seems unavoidable, a *hard braking* is applied. Also, a pre-fill of the brake circuit in case of possible risk (when the FCW system is launched) can be provided, in order to be ready for a full-brake in case it is required (either by the driver or automatically). In case the system is not able to avoid an accident but can help in the collision mitigation as the

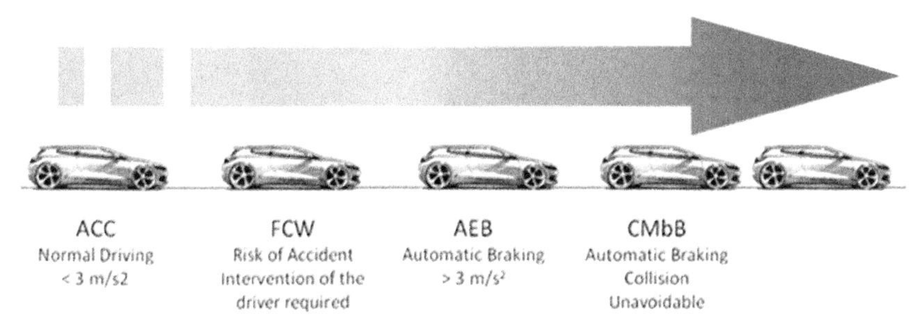

Figure 9.2 Stages on the longitudinal control of the vehicle.

obstacle is crashed at a lower speed, it is called CMbB, Collision Mitigation by Braking. The only difference is that AEB can really avoid the accident, while CMbB is launched a short time before the accident that can't be avoided any more.

SLA (Speed Limit Assistant)

The Speed Limit Assistant (SLA) is a safety system that provides the driver with information on the most suitable maximum speed continuously during his or her journey.

SLA system can be based on several sub-systems:

- TSR (Traffic Sign Recognition): Recognition of the traffic signs on the road, either by vision or gathering information from a map, is shown to the driver as a reminder of the prevailing speed limits.
- CSW (Curve Speed Warning): As extracted from the digital maps, information of the most suitable recommended maximum speed limits

Figure 9.3 CSW system.

Figure 9.4 TSR system.

when passing the curve ahead are shown to the driver. Another option is to show just a warning icon in case speed is considered as too high for the incoming bend.

9.2.2 Lateral Control Systems

Lateral control systems take care of the lateral dynamics of the vehicle, either warning the driver or taking control of the vehicle actuation systems.

LDW/LKA (Lane Departure Warning/Lane Keeping Assistant)

The Lane Departure Warning system has the task to warn the driver in case he drives out of the lane due to a distraction (without using the blinkers). Many OEMs offer today a Lane Departure System under different commercial brands (AFIL, Audi Lane Assist, etc.). It is composed by a sensor (or several sensors) with the capability to detect when the driver is leaving from the chosen lane, a Control Unit and a suitable HMI for the driver.

Lane's lines detection can be done through two different technologies:

- Infrared sensors placed in the low part of the vehicle (PSA models): They use the reflection produced by the emitted light when driving over a white line to detect if the vehicle is driving over them. In this case, a Control Unit determines the driver is departing from the lane, and, depending on some other factors (blinkers, etc.), it can warn him or her by different methods (making the steering wheel or the seat vibrate, sound warning, etc.).
- Image processing: A camera—usually placed behind the windshield, on the rear view mirror housing—provides images which can be analyzed. Thus, it is possible to determine when the driver is departing from its chosen lane. This system brings advantages, such as its predictive capability (it can on obstacles in the already known driving corridor) and is more robust in front of situations such as arrows, providing considerably fewer false alarms. As a disadvantage, it can be less robust in case of poor visibility.

In any case, the system works from a certain speed (commonly, from in between 60 and 80 km/h upwards) and can be switched off. Moreover, when activating the suitable blinker, the system understands that the driver really wants to change lane and no warning is provided in case of crossing the lines.

An update of the system is also found in the market: LKA (Lane Keeping Assistant), which includes an additional torque on the steering wheel (electrical

Figure 9.5 LDW system.

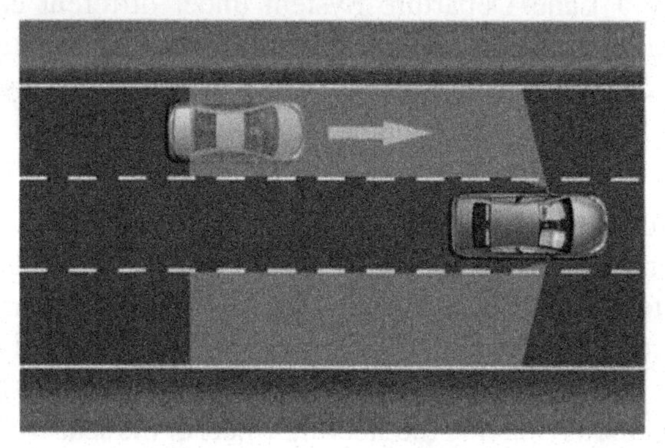

Figure 9.6 BSD/LCA system.

power steering is required) that helps the driver to keep the vehicle into the desired lane.

BSD (Blind Spot Detection)

A Blind Spot Detection system has the goal to warn the driver in case another vehicle is located in the blind spot which is not controlled by the rear-view mirrors.

Therefore, it counts on some sensors (commonly, short range radars @ 24 GHz or image processing units) which monitor constantly the area placed in the lateral blind spots of the vehicle. These sensors provide information to a Control Unit, which decides the susceptibility to provide the driver with a warning. This warning can be acoustic, visual or haptic.

Some systems can warn continuously on the existence of objects in the blind spot. Some others only warn when the driver expresses his or her will to change lane, using the correspondent blinker. They usually work over a certain speed and are capable to exclude parked vehicles or those driving in the opposite direction, in order to reduce the false alarm rate. The detection area can measure around 10 meters behind the rear view mirror and 4 meters wide, enough to cover the blind spot.

LCA (Lane Change Assistant)

A Lane Change Assistant is a system which increases the possibilities of a Blind Spot Detection System. The detection distance can achieve up to 50–60 meters behind the ego-vehicle (positions and speed profile of the vehicle) in the adjacent lanes. Moreover, the relative speed of the detected vehicles is also taken into account, so the system is capable to warn the driver in case the lane change is too risky because of a fast approaching vehicle from behind. Depending on some parameters, different warning levels can be included.

9.2.3 Other Control Systems

Pedestrian detection/Active hood

A pedestrian detection system is capable to recognize a potential danger. In this case, the driver can be warned or even an automatic action can be performed (automatic speed adaptation). In case of unavoidable crash, the activation of passive safety measures is also considered (active hood).

PreCrash systems

In the transition or overlap between active and passive safety, PreCrash systems work when accidents are unavoidable. Its mission is, based on the information gathered by the rest of the safety systems, and after determining the accident cannot be avoided by its intervention, to prepare the passive safety elements of the vehicle to better perform their safety mission. For instance, when there's a sure head-on collision, CMbB will reduce the speed of the crash, while PreCrash will pre-tension the seatbelts, will move the seats to place them in a more convenient position or will pre-trigger airbag deployment order. PreCrash systems can cover the front of the vehicle, the rear or all 360° of the vehicle.

Parking assistance

Parking assistance is one of the most implemented DAS. There are many types of technology used on this. This section will not be focused on the traditional ultrasonic or vision aided parking assistance systems, but on the systems that can provide some kind of support to the driver. These systems can be divided in the following ones:

- *Vision-Aided Systems*: together with the image of a camera placed in the rear part of the vehicle, some support provided by visual guidelines in the dashboard display.
- *Top View Systems*: up to 4 cameras placed on exposed surfaces around the vehicle provide images that, after some processing, can be shown on the vehicle's display as if it was seen from above.
- *Aided Park Systems*: some systems can provide support to the driver on his/her search for parking spots or his/her maneuvers to park the vehicle.

Figure 9.7 Top view of a parking assistance system.

Figure 9.8 Aided park system.

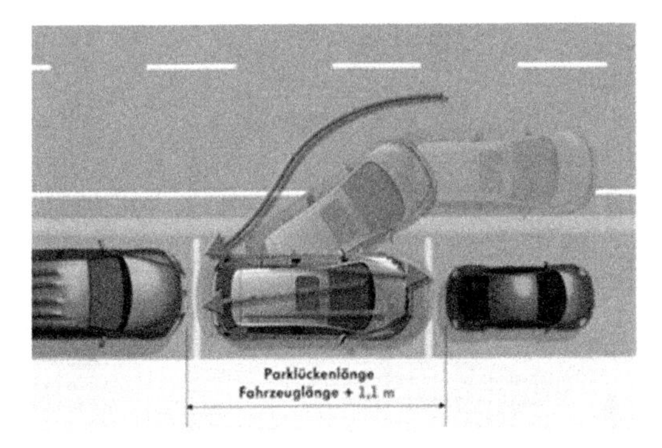

Figure 9.9 Automatic park systems.

- *Automatic Park System*: this system can take control of the steering of the vehicle in order to park automatically after detection of a parking slot. The driver remains responsible for the longitudinal control of the vehicle.

9.2.4 Control Solution in ADAS

Based on most control architectures for Automated and semi-automated vehicles [2], DESERVE is divided in three main platform parts or stages: perception, application and information-warning-intervention (IWI). The sensing and perception of environmental and onboard information is vitally important for any automotive DAS function. Based on preliminary work from other funding projects in this area[3] the information flow and architectural decomposition of the DESERVE platform is shown in Figure 9.10.

The three main building blocks in Figure 9.10 are the perception layer, the application layer and the IWI controller layer. The same decomposition was also chosen from other parties in similar projects (like InteractIVe [3]) and corresponds to the naturalistic behavior that is applied when accomplishing a given task, namely the action points "sense", "plan" and "act". As baseline DESERVE considers the results of several research projects, like InteractIVe, but targets the standardization of the software architecture.

Indeed, by handling the sensor and actuator information on a virtual and abstract level, a systematical standardization of input and output interfaces can be realized. This results both in a very good encapsulated module architecture and makes exchange or addition of further module components much easier.

[3]InteractIVe—FP7/ICT funding project—www.interactIVe-ip.eu

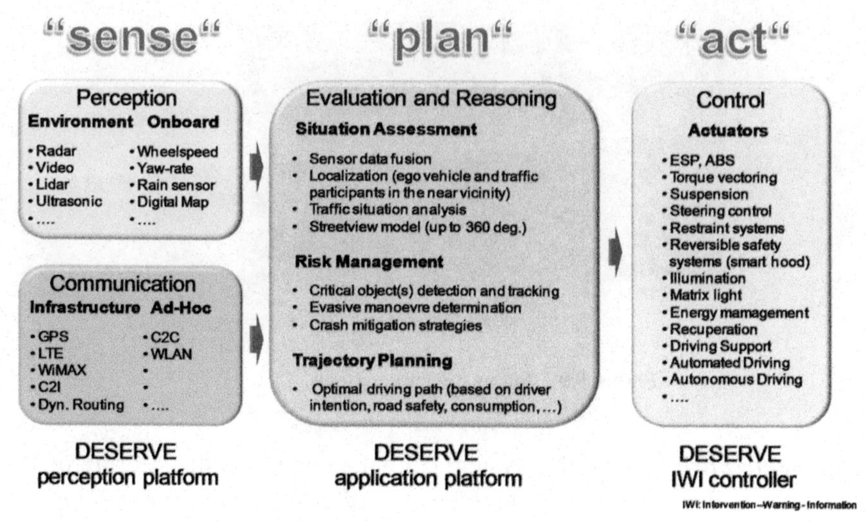

Figure 9.10 DESERVE platform.

In particular, the Perception Platform processes the data received from the sensors that are available on the ego vehicle and sends them to the Application Platform. The data received from the Application Platform are used to develop control functions and to decide the actuation strategies. Finally, the output is sent to the IWI Platform informing the driver in case of warning conditions and activating the systems related to the longitudinal and/or lateral dynamics.

9.2.4.1 Perception platform

The main objective of the Perception layer is to define and develop the DESERVE platform components that will interface with sensors and actuators, acquiring information from the typical sources. All these possible information sources are addressed, described and characterized in an abstract level that allows virtualization of input and output data. By using such an abstract and virtual intermediate layer the connection/exchange of sensors or actuators and the porting or adaptation to different vehicle models is expected to become much easier and less time consuming.

The DESERVE Perception layer is composed of different sub-layers that build up, in their totality, the complete information source that can be imported into the DESERVE platform framework. In a generalized sense the Perception layer can be seen as the input and output (I/O) gateway, especially when including communication devices and the different actuators as part of the I/O components.

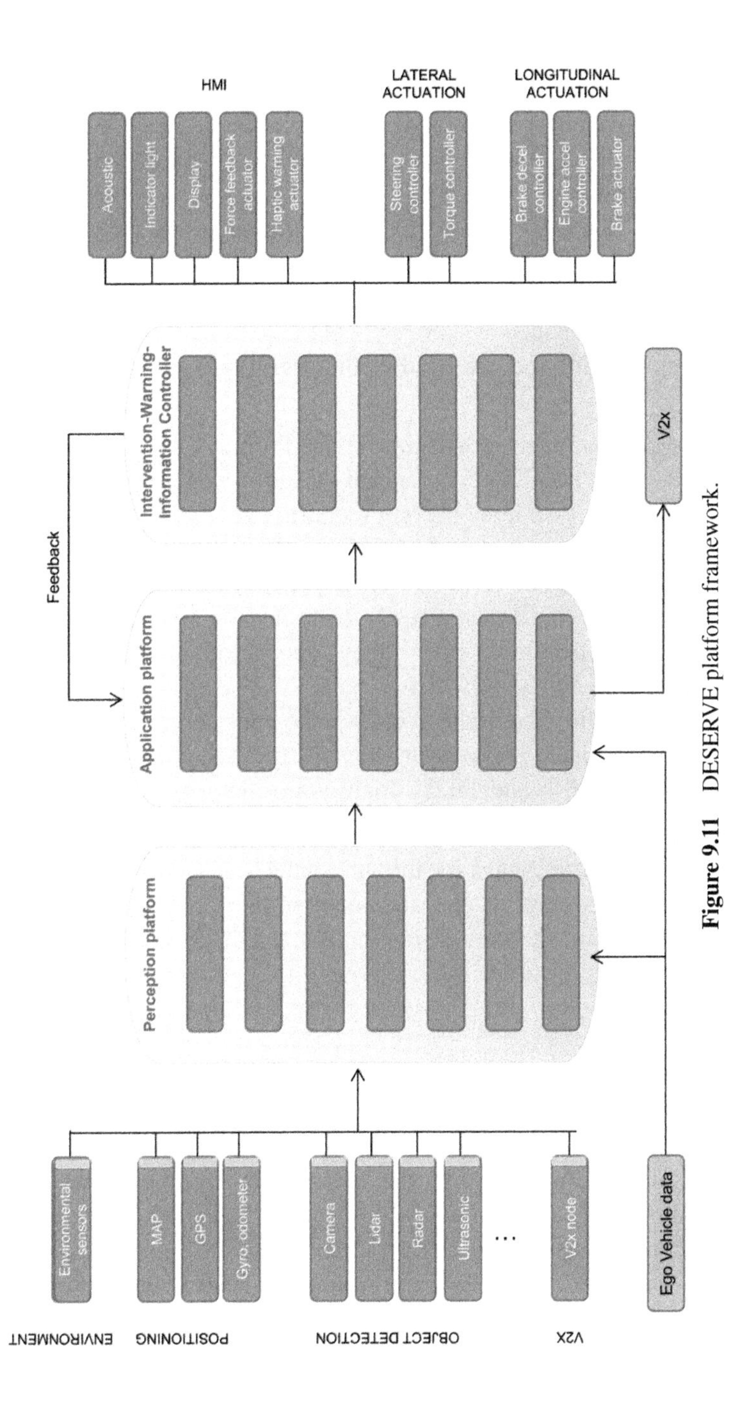

Figure 9.11 DESERVE platform framework.

9.2.4.2 Application platform

Based on these assumptions and previous works, a control strategy for sharing vehicle control between the driver and embedded ADAS systems was proposed. These layers can be used dynamically, based on the information from the driver monitoring automotive—DMA.

Since the driver is legally responsible for operating the car in its environment, in our approach he/she will have the last responsibility in the arbitration control process. However, if the driver is not enabled to drive, then the control will be taken by the embedded system.

The specific Application modules used in the arbitration and control of the vehicle are:

- Threat assessment: the information from Frontal Object Perception, Vehicle trajectory and Driver intention modules will be considered, in order to establish a risk level in each scenario.
- IWI manager: this module will determine the action to be taken by the driver or the vehicle (here we can set the Arbitration and Control functions). The Driver Assistance Systems involve two main decision makers: when is the driver who takes the control or when does the automated system and up to which extent.
- Vehicle control: Only the brake pedal will be considered. Classical control techniques considering comfortable/safe accelerations. Longitudinal control based on PID and Fuzzy logic controllers have been used in automated functions.

The level of assistance provided by the automated car to the driver might change depending on the driver's state and on the situation at hand (imminence of danger). With a varying level of automation of the automated vehicle, control might smoothly flow from the driver to the automated car and vice versa.

9.2.4.3 Information Warning Intervention (IWI) platform

The Information Warning and Intervention module uses the output of the Application layer and provides ways to execute the interaction with the driver and the control of the vehicle. Mainly the information is sent to the actuators that will translate high level commands into acceleration and steering angle to provide the correct answer expected from the vehicle.

In a similar way, information is sent through the HMI towards the driver if necessary. These messages will warn and inform the driver (visual and acoustic signals/messages), as well as interact with him/her (haptic signals).

In order for these messages to be effective, great efforts have been done in HMI solutions where the current hot topic is to share the control with the driver. In the following, a review of some techniques for the arbitration and shared control are presented.

9.3 Survey on Arbitration and Control Solutions in ADAS

In the transportation field, human machine interaction plays a key role. Nowadays, significant results have been achieved in the automated driving field (at least, under certain circumstances) [4, 5]. Nonetheless, there is a long way to go before removing the driver from the loop in real traffic conditions.

Parasuraman et al. [7], stated that the main problem in this kind of systems lies in the decision making process and the assignment of control responsibility. In the ITS field, shared control is the action of carrying a task simultaneously between a (on board) computer and a driver, differing from manual control and fully automation (since no real "sharing" is being done in this situations, see Figure 9.12).

The first levels of automation were set by Sheridan in [9]. Here, 10 different levels described the amount of responsibility for each decision maker. Flemisch et al. in [10] presents a more developed view of the levels needed for control sharing, where the automation is based in the H-metaphor and clarified in two main groups: Tight rein and loose rein.

Recently [11], new taxonomy of automated driving was issued by SAE International; its control levels are depicted by Figure 9.12. Other levels of automation have already been proposed by the German Federal Highway Research Institute (BASt) [12] and the National Highway Traffic Safety Administration (NHTSA) [13]. A comparison of these is summarized in [11], stating that the SAE taxonomy is alike the other two, but gives a broader and more specified view of automation levels. For this reason, the SAE taxonomy will be the one taken into account (see Figure 9.12).

When considering the driver in the control loop, it is important to know the automation level embedded in the vehicle. This will permit the control

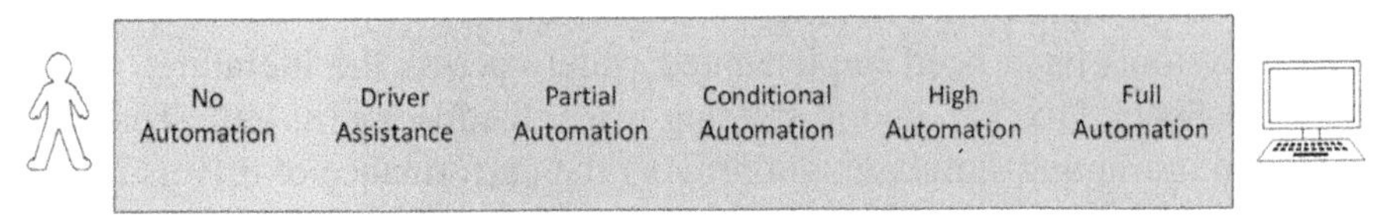

Figure 9.12 SAE J3016 standards of driving automation levels for on-road vehicles.

sharing system to set the limits for each decision maker. We will deepen in the arbitration concept as a way to change, in a smooth way, the level of control according to the situation in-hand.

The **Arbitration concept** is *the process of settling an argument or a disagreement by an entity that is not involved.*[4] Little research has been done in terms of arbitration (since it is a new concept in vehicle automation). First approaches define cognitive states and relations between humans and machines [6], also mental models as in human relationships have been considered by [14]. This consideration leads to a scenario where the status of the driver and the system must be known, at all times, aiming to set an accurate level of automation for the current situation.

From the above, communication between the system and the driver should constantly occur, in a way that is possible for both to make a mental model of one another [14]. Also different metaphors have been stated, such as the copilot metaphor (referring to the automated system) and the H-metaphor as a comparison between horse-human cooperation and vehicle-human cooperation [15].

9.4 Human-Vehicle Interaction

Increasing need to pay more attention to the human driver in interaction with the vehicle has been recently identified [1]. From other domains where automation is already widely used (e.g. aviation, central rooms) it is known that automation has both positive and negative effects on the human operator. With increasing automation in the vehicle domain these effects need to get far more attention on the short term, evaluating the human-vehicle relationship and assigning countermeasures if necessary [1]. In order to have a regular communication between the two decision makers (the driver and the embedded system), in [15], a haptic HMI system is proposed where active force feedback is the common language. This allows the message to be directly linked to the actuator where the reaction of the driver is expected, also allowing the system to evaluate the performance of the driver. The haptic feedback can also give hints in terms of the action the driver should perform (e.g. the steering wheel turns a little to the right or left in order to hint the driver).

Haptic systems have been implemented widely across the literature: in gas pedal feedback [16, 17], and in steering wheel feedback [18, 19]. These are also used in training simulators, improving the performance of drivers in different scenarios.

---

[4] Oxford dictionary.

The use of corrective feedbacks is known to cause over-corrective behavior [8] or bad performance when removed. This happens because it impairs the input-output relationship in motor skill learning of the driver. In [20], the haptic aid shows a good performance if the feedback is provided as needed and not all the time.

For arbitration and shared control, a state of the driver is needed in order to know his current status to perform the driving task. In [21], an extensive study on driver distraction was performed. It showed that in terms of visual and cognitive attention sharing, while performing following or passing driving maneuvers, a warning from the HMI proved to be helpful.

In [22], the importance of vision at the driving task was stated. Although visual acuity proved to be important, other indicators of the driver ability (Visual field, processing speed, divided attention, among others) have evidence-basis for their relevance to the driver ability and safety, and can be measured in a noninvasive way with recent in-car perception systems, as in [23].

Recently, the HAVEit[5] project [24, 25], and the InteractIVe[6] project [26] have made the first approaches into control sharing strategies, theoretically and in simulations, with driver-in-the-loop capabilities.

The aim of arbitration and control solutions in ADAS, inside the DESERVE project is to effectively share the control with the driver and manage risky situations. In [27], ADAS applications are listed such as lane change assistance systems, pedestrian safety systems, adaptive light control, and parking assistance systems, among others. These are considered to improve the automated system and take into account the driver-in-the-loop for arbitration applications [28].

Arbitration systems for shared control applications is a new concept in the ITS research field. Based on previous contributions, it is the objective to develop a system able to share the control—in a smooth way—between the decision makers. Motivation for this approach can be found in social needs [29], legal challenges [1, 33] and technical bases such as the DESERVE platform (see [11]).

9.5 Driver Monitoring

Driver's limitations are very often related to his physiological and psychological states. An optimum pilot state includes an optimum alertness level

[5]http://haveit-eu.org/

[6]http://www.interactive-ip.eu/

and a task-oriented attentiveness. The distinction between "alertness" and "attention" is justified in the way that driver "alertness" is presumed to be necessary but not sufficient for an appropriate focus on external events. Thus, drivers may be alert but still be inattentive. In order to assess alertness and attentiveness in the DESERVE project, two main factors are evaluated:

- Drowsiness/fatigue
- Distraction

Up to now, a universally valid definition of drowsiness still lacks. A tired driver mainly derives from performing a highly demanding task for extensive time periods ("time-on task" for the driving effort). Other definitions focus on the sleepiness level, which is the state of being ready to fall asleep. It is mainly caused by circadian rhythms and sleep disorders (reduced quality or quantity of sleep).

On the other hand, *"Driver distraction refers to those instances when a driver's attention is diverted from the primary task of driving the vehicle in a way that compromises safe driving performance"*, [30]. This distraction can be either internal (e.g. other passengers interaction, cellphone, etc.) or external (e.g. other road users, traffic signs, etc.). It can also be classified in different modes as: Visual (external attractors for example advertisement on the side of the road or internal attractors e.g. looking to his children at the back of the vehicle, displaying an address onto a navigation device, etc.), acoustic (ringing phone, listening music) or cognitive distraction (conversing at phone but also internal thought and rumination, etc.).

For more information about on-line driver monitoring approaches, the reader is referred to [34]. Here a description of the different on-the-market and research methods and approaches are described in detail. In the DESERVE project, two main approaches were taken into consideration for the assessment of alertness and attentiveness of the driver:

The Continental driver supervision system is implemented for a real time monitoring of two independent parameters, the drowsiness level (sleepiness vs. awakeness) and the visual inattention (e.g. the driver "is/is not" looking to the road) [23].

The Driver state monitoring includes a compact low consumption and high dynamic range (120 dB) CMOS camera sensor. The camera is equipped with a global shutter for the synchronization with a set of pulsed NIR lights (850 nm).

Ficosa's Somnoalert Sensor aims to detect "non-apt to drive" states using physiological signals such as thoracic effort signal. An external thoracic effort sensor sends the signals to a smartphone, where it is processed to evaluate the state of the driver and indicate if this becomes dangerous.

9.5.1 Legal and Liability Aspects

For automated vehicles, it is still unclear how legal and liability aspects are going to evolve. As a matter of fact, the U.S. legislation does not prohibit nor allows the use of automation in the driving task [31]. This leaves an important legal gap towards the responsibility of any action taken by the on-board system, since it is now an entity that "thinks for itself". Similar situations arise in Europe where in a crash the responsible at all times is the driver, even when an embedded system was controlling the vehicle [32].

From the legal perspective, several initiatives in the U.S., specifically in the states of Nevada (2011), Florida (2012), California (2012), Washington D.C. (2012) and Michigan (2014), have already established some of the minimum safety requirements in order to allow automated vehicles technology [33]. Other state legislations in the U.S. are following these initiatives, to take a wider view of this the reader is referred to [32] and [33]. In the E.U., initiatives launched between governments and manufactures are currently creating the framework for the new standards and regulations for automated driving. These address legal matters and promote the standardization of the automated vehicles technology, as for example the Citymobil2 project [36].

As to liability, Beiker and Calo [35] noted that the situation is more complex with automated vehicles, concluding that it is unclear how the courts, or the public, will respond to the prospect of artificial intelligence acting on behalf of humans with fatal consequences. They expect that a set of policies can be established to create the necessary legal framework for further development of vehicle automation. In the E.U., the legal framework sets the liability of any crash towards the driver. This creates many barriers for automated vehicles and restricts them to private roads.

As a matter of fact, automation (or the lack of it) is not black or white but rather in shades of gray, complex and involving many design dimensions [1]. OEMs are careful with this and do not claim that an ADAS is working in all driving situations. A helpful model of automation is to consider different levels of assistance and automation that can e.g. be organized on a scale as in [11]. This not only suggests but encourages the use of systems that consider the driver-in-the-loop. These systems will allow the industry to add driver's vigilance to their system's supervision and avoid gaps (at least in the legal framework).

9.6 Sharing and Arbitration Strategies: DESERVE Approach

The arbitration module is defined in the information, warning, intervention (IWI) manager (Application platform) of the DESERVE abstraction layer (Figure 9.11). This Advanced Driver Assistance System involves two main decision makers: the driver and the automated system. It will determine the level of responsibility of each of them at all times and allow smooth transitions between automation levels defined in [11].

Based on the information from different perception systems, it is possible to define fuzzy control parameters to achieve this, as was proposed in [37, 38]. This cognitive process will result in the selection of a course of action among several alternative scenarios (e.g., up to which amount the driver should be responsible of the pedal action in an ACC maneuver while tired). The proposed system consists of a two level fuzzy approach for the arbitration (IWI manager) and vehicle sharing (VMC) modules.

The arbitration and sharing control concept has been developed in RTMaps, one of the development platform defined in DESERVE. Figure 9.13 shows the general diagram for the arbitration. Here a fuzzy logic approach is implemented to compute the automation assessment (or situation status of decision-makers). This value is an assessment of the alertness and attentiveness of the driver w.r.t. the risk detected from the situation status.

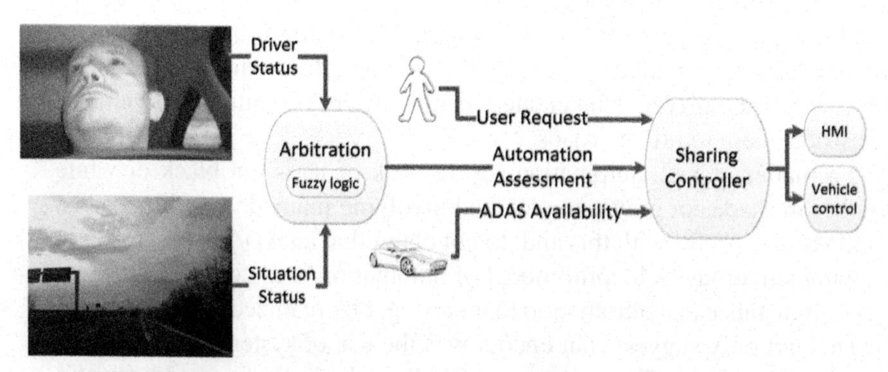

Figure 9.13 Arbitration and control sharing application: General diagram.

The sharing controller considers the automation assessment (but also the driver and the automated systems decisions) to decide the level of control and responsibility of each decision maker in real time. The output goes then to the HMI, informing the driver (a haptic steering wheel system informs the driver of next maneuvers that the system is ready to perform), and to the vehicle control. This process is done in real time, allowing a smooth sharing between decision-makers. For details and further perspective in first preliminary results please refer to [38].

9.7 Conclusions

This chapter presents a survey on arbitration and control solutions for ADAS, based on the ADAS solutions available in the market, and the ones considered from the functional requirements described in Sub-Project-1 of the DESERVE project. The main architecture is described as a three-pillar platform system first "sensing" the environment, then "planning" according to decisions made over perception data and finally "acting" to follow those decisions.

For the sharing and arbitration approach, different points of view have been considered. Here, the estimation of the driver state and the assessment of the risk related to the situation in hand are the most important ones. These allow the system to have a coherent evaluation of the situation of both decision makers and arbitrate if the vehicle's embedded system needs to intervene because of risky driver actions.

This intervention is performed through haptic signals. However, there are still some challenges with respect to HMI solutions that can properly work as a communication bridge for the two decision makers and inform the driver—on time—of automated vehicles decisions.

Furthermore, legal and liability aspects are important milestones yet to be tackled. Although some states of the U.S. are taking the initiative, law regarding automated vehicles is in its first steps. Liability and legal responsibility still lies with the driver, hence, in our approach the control lies with the drivers (the driver can deactivate the system at any stage and is stronger than haptic cues). In future research we will focus in the arbitration, to determine (using some perception information) up to which point the embedded system can take control of the vehicle and which situations are more dangerous (risk management, taking special care of situations where overreliance on the system occurs—the embedded system returns the control to the human driver).

References

[1] Schijndel-de Nooij, Margriet, Krosse, Bastiaan, Broek, Thijs, Sander Maas, Ellen van Nunen, and Han Zwijnenberg. "*Definition of necessary vehicle and infrastructure systems for Automated Driving*", Study Report for the European Commission (2011).

[2] González David, Pérez Joshué, Milanés Vicente and Nashashibi, Fawzi., "*A Review of Motion Planning Techniques for Automated Vehicles,*" in Intelligent Transportation Systems, IEEE Transactions on, in press.

[3] Hesse, Tobias, et al. *Towards user-centred development of integrated information, warning, and intervention strategies for multiple ADAS in the EU project interactive.* s.l.: Universal Access in Human-Computer Interaction. Context Diversity, 2011.

[4] Ziegler, Jens, Philipp Bender, Markus Schreiber, Henning Lategahn, Tobias Strauss, Christoph Stiller, Thao Dang et al. "*Making bertha drive—An autonomous journey on a historic route.*" Intelligent Transportation Systems Magazine, IEEE 6, no. 2 (2014): 8–20.

[5] Broggi, Alberto, Pietro Cerri, Stefano Debattisti, Maria Chiara Laghi, Paolo Medici, Matteo Panciroli, and Antonio Prioletti. "PROUD-Public road urban driverless test: Architecture and results." In *Intelligent Vehicles Symposium Proceedings, 2014 IEEE*, pp. 648–654. IEEE, 2014.

[6] Hoc, Jean-Michel. *Towards a cognitive approach to human-machine cooperation in dynamic situations.* s.l.: International Journal of Human-Computer Studies, 2001. Vol. 54.

[7] Parasuraman, Raja, Sheridan, Thomas B and Wickens, Christopher D. *A Model for types and levels of humans interaction with automation.* s.l.: Systems, Man and Cybernetics, Part A: Systems and Humans, IEEE Transactions on, 2000. Vol. 30.

[8] Leeuwen, Peter van, et al. *Effects of concurrent continuous visual feedback on learning the lane keeping task.* s.l.: Proceedings of the Sixth International Driving Symposium on Human Factors in Driver Assessment, Training and Vehicle Design, 2011.

[9] Sheridan, Thomas B. and Verplank, William L. *Human and Computer Control of Undersea Teleoperators.* s.l.: Massachusetts Institute of Technology Cambridge Man-Machine Systems Lab, 1978.

[10] Flemisch, Frank, et al. *Cooperative Control and active interfaces for vehicle assisstance and automation.* s.l.: FISITA World automotive Congress, 2008.

[11] Taxonomy and Definitions for Terms Related to On-Road Motor Vehicle Automated Driving Systems. s.l.: SAE International, 2014.

[12] Gasser, Tom M. and Westhoff, Daniel. *BASt-study: Definitions of Automation and Legal Issues in Germany.* s.l.: 2012 Road Vehicle Automation Workshop, July 25, 2012.

[13] *Preliminary Statement of Policy Concerning Automated Vehicles.* s.l.: Nationam Highway Traffic Safety Administration, May 30, 2013.

[14] Flemisch, Frank, et al. Automation Spectrum, inner/outer compatibility and other potentially usefull human factors concepts for assisstance and automation. s.l.: Human Factors for assisstance and automation, 2008.

[15] Flemisch, Frank O., et al. *The H-Metaphor as a Guideline for Vehicle Automation and Interaction.* s.l.: NASA Center for AeroSpace Information, 2003.

[16] Abbink, David A. *Neuromuscular analysis of haptic gas pedal feedback during car following.* s.l.: Faculty of Mechanical Maritime and Materials Engineering, Delf University of Technology, 2006.

[17] Winter, Joost C.F. de, et al. *A two-dimensional weighting function for a driver assistance system.* s.l.: Systems, Man, and Cybernetics, Part B: Cybernetics IEEE Transactions on, 2008. Vol. 38.

[18] Mulder, Mark, Abbink, David A. and Boer, Erwin R. *The effect of haptic guidance on curve negotiation behavior of young, experienced drivers.* s.l.: Systems, Man and Cybernetics, 2008 SMC 2008 IEEE International Conference on, 2008.

[19] Abbink, David A., Mulder, Mark and Boer, Erwin R. *Haptic shared control: smoothly shifting control authority?* s.l.: Cognition, Technology & Work, 2012.

[20] Crespo, Laura Marchal, et al. The effect of haptic guidance, aging, and initial skill level on motor learning of a steering task. s.l.: Experimental Brain Research, 2010. Vol. 201.

[21] Zhang, Yu. Visual and Cognitive Distraction Effects on Driver Behavior and an Approach to Distraction State Classification. Raleigh, North Carolina: North Carolina State University, 2011.

[22] Owsley, Cynthia and Jr., Gerald McGwin. *Vision ans Drivng.* s.l.: Vision Research, 2010. Vol. 50.

[23] Boverie, S., Cour, M and Le Gall, JY. *Adapted Human Machine Interaction concept for Driver Assistance Systems DrivEasy.* Milano: 18th IFAC World Congress, 2011.

[24] Flemisch, Franck, et al. *Towards Highly Automated Driving: Intermediate report on the HAVEit-Joint System.* Brussels: 3rd European Road Transport Research Arena, Tra2010, 2010.

[25] Vanholme, Benoit. Highly Automated Driving on Highways based on Legal Safety, PhD Thesis. 2012.

[26] Hesse, Tobias, et al. *Towards user-centred development of integrated information, warning, and intervention strategies for multiple ADAS in the EU project interactive.* s.l.: Universal Access in Human-Computer Interaction. Context Diversity, 2011.

[27] *D24.1 -Vehicle Control Solutions-, SP2.* s.l.: DESERVE project, 2013. Deliverable.

[28] D44.1 *-Automated Functions Solution Design-*, SP4. s.l.: DESERVE project, 2013. Deliverable.

[29] Adams, Lisa D. *Review of the literature on obstacle avoidance maneuvers: braking versus steering.* s.l.: Univ. Michigan Transp. Res. Inst., Ann Arbor, MI, Tech. Rep. UMTRI-94-19, 1994.

[30] Young, Kristie, John D. Lee, and Michael A. Regan, eds. Driver distraction: Theory, effects, and mitigation. CRC Press, 2008.

[31] Walker Smith, B. *Automated Vehicles Are Probably Legal in the United States,* 1 Tex. A&M L. Rev. 411, 2014.

[32] Trimble, Tammy E., Richard Bishop, Justin F. Morgan, and Myra Blanco. *Human factors evaluation of level 2 and level 3 automated driving concepts: Past research, state of automation technology, and emerging system concepts.* No. DOT HS 812 043. 2014.

[33] Anderson, James M., Kalra Nidhi, Karlyn D. Stanley, Paul Sorensen, Constantine Samaras, and Oluwatobi A. Oluwatola. *Autonomous vehicle technology: A guide for policymakers.* Rand Corporation, 2014.

[34] D32.1 *-General Driving Monitoring module definition SoA-*, SP3 s.l.: DESERVE project, 2013 Deliverable.

[35] Beiker, S. and Calo, R. *Legal aspects of autonomous driving.* Santa Clara L. Rev. 52 (2012): 1145.

[36] J. van Dijke and M. van Schijndel, *Citymobil, advanced transport for the urban environment: Update,* Transportation Research Record: Journal of the Transportation Research Board, no. 2324, pp. 29–36, 2012.

[37] D24.4 *-Generic ADAS Control-*, SP2 s.l.: DESERVE project, 2013 Deliverable.

[38] Perez, J. M., David Gonzalez, Fawzi Nashashibi, Gwenael Dunand, Fabio Tango, Nereo Pallaro, and Andre Rolfsmeier. “Development and design of a platform for arbitration and sharing control applications.” In Embedded Computer Systems: Architectures, Modeling, and Simulation (SAMOS XIV), 2014 International Conference on, pp. 322–328. IEEE, 2014.

PART III

Validation and Evaluation

10

The HMI of Preventing Warning Systems: The DESERVE Approach

Caterina Calefato[1], Chiara Ferrarini[1], Elisa Landini[2], Roberto Montanari[2], Fabio Tango[3], Marga Sáez Tort[4] and Eva M. García Quinteiro[4]

[1]Unimore – University of Modena and Reggio Emilia – Italy
[2]RE:Lab srl, Italy
[3]CRF – Centro Ricerche Fiat, Italy
[4]CTAG – Centro Tecnológico de Automoción de Galicia, Spain

10.1 Introduction

Early '70s literature in traffic safety put into evidence how the majority of accidents is a consequence of human error. One of the pioneering work carried out in 1977 in the automotive domain [34] started from an examination of a large number of accidents and showed that more than 90% of them was determined by different kind of mistakes attributable solely to a human factor and rarely to technical and/or environmental failures.

This finding was confirmed in the following years also in other domains with very complex technologic contexts (i.e. avionic, railway, etc.).

It was realized that in the framework of the evolution of technical systems, the human element plays a fundamental role both as a governing factor and as a potential menace to safety. This concept paved the way for the modern preventive safety systems, wide known as ADAS (Advanced Driver Assistance System).

The experience carried out into the DESERVE project (Development Platform for Safe and Efficient Drive) was agreed by all involved partners to be beneficial for the extension of future ADAS. A key role in this process is played by the Human Machine Interface (HMI). Since ADAS systems cope with the driving task influencing driver's decisions or directly intervening in the driving maneuver, the issue of the driver's trust opens a crucial design problem, because the driver cedes a part of the control [30]. Low trust, resulting e.g. from an earlier experience of failure, can lead to disuse of the system [24]. Building and enforcing the driver's trust through a positive system experiencing depends not only on the proper functioning of the system itself (i.e. the capability of detecting some events) but also on the HMI design.

In order to create those positive experiences and avoid the disuse of ADAS, one has to understand the driver and his/her goals and motives while driving [13], together with the role of technology in supporting the driver in his/her task and in avoiding road accidents.

This chapter aims at exploring step by step the rationale behind the effective design of the Human Machine Interface for ADAS systems, giving the reader an outline of the role and scope of ADAS system. In next paragraphs, a particular focus on the role of humans and role of technology in the preventing of the road accidents is presented, along with the discussion of the importance of the detection of the driver's intention. Then an example of a whole HMI design process is presented. In fact during the DESERVE project the in-vehicle HMI for 17 functions (13 of them were ADAS) was designed and evaluated. This chapter will report to the reader how the HMI was conceived, including discussions on the role of ADAS in preventing imminent accidents and a short state of art on HMI design approaches.

10.2 Prevent Imminent Accidents: The Role of Humans, the Role of Technology

In general, the amount of accidents among the years is progressively decreased since the second half of the '80s [8]. This basically depends both from a strengthen in humans awareness on the accident causes, partly influenced by the evolution of studies in humans factors, but mostly this depends on technological innovation on vehicles. The history of such evolution which intends to show the relationship existing between the run-up of the accident and the technologies and functions for safety enhancement will be presented in the next paragraph.

10.2.1 From Passive to Preventive Safety

The first phase in reaching a higher safety degree on vehicles was due to the introduction of the so called passive safety systems, whose main purpose is to improve the driver conditions while an accident takes place. Indeed, the

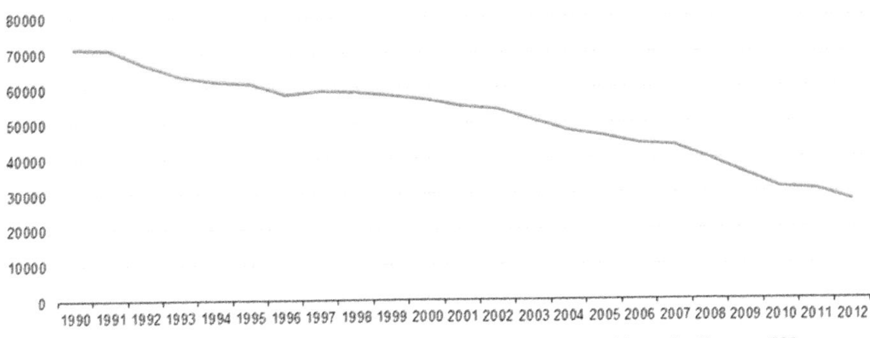

Figure 10.1 Total number of fatalities in road traffic accidents in Europe [8].

introduction of safety belts, airbag, etc., as well as the strengthening of the materials have significantly reduced the number of injuries and consequently the number of victims on the road. For instance, studies on the effectiveness of the seat belts were conducted since the end of the '60s starting from Sweden [4].

The second phase was characterized by the introduction of active safety systems, which were intended to increase the safety of the driver when approaching a dangerous situation. In particular, this period dealt with the introduction of systems such as the ABS (Anti-lock Braking Systems), the ESC (Electronic Stability Control), as well as other functions able to intervene by minimising the impact in proximity of a potential dangerous situation and, hence, by avoiding the accident. For instance cars equipped with ESC were 22% less likely to be involved in crashes than those without, with 32% and 38% fewer crashes in wet and snowy conditions [19].

The challenge of reducing even more the number of accidents consists in allowing the development of the so called *preventive safety technology*, which is conceived to assist the driver when the risk of occurring a hazardous and critic situation is greeting higher. These technologies, named ADAS (Advance Driver Assistance Systems) are able to monitor the driving dynamics by introducing preventive features in support of the driving activity. In particular, driving safety will be fostered on the longitudinal axis of the vehicle thanks, for instance, to the frontal collision warning and adaptive cruise control systems. Driving safety on the lateral axis can be improved by systems like lane support and lane warning. The implementation of blind spot improves the safety on the rear spectrum indeed. The purpose of this approach is twofold: on one hand it is intended to guarantee an high level of protection on the road, almost as if the driver was stuck inside of a kind of "safety bubble", as highlighted by some researchers when referring to the concept of "virtual safety belt" [31]. On the other hand, it aims at allowing cars to operate in coordination by implementing a scenario where the whole vehicles have high situation awareness capabilities. It is indubitable the effectiveness of the ADAS in driving safety, even most of them have not yet achieved a mature introduction in vehicle market but are still in the prototyping phase. Nevertheless, researches has shown that to an increase of the automation and accident prevention features included in the on-board technologies does not always correspond to increase of the driving confidence, especially if the drivers' expectations in vehicle technologies interaction are not fully taken into consideration by designers. On the other hand, a theory known as Peltzman effect [25] seems to show that an improvement of confidence due to effective automated safety support systems, even if they are only able to increase the driving monitoring scenario, could induce drivers in improving, for instance, speed, till to jeopardize the effectiveness of such systems.

10.2.2 The Role of Driver Model in ADAS Design

As aforementioned, Advanced Driver Assistance Systems (ADAS) have been implemented more and more in recent years in the automotive industry, in order to move from passive safety to preventive safety. In this context, through the driver models, a more complete understanding of driver's behaviour is expected to have the opportunity to enhance the road safety and to increase the driver acceptance of in-vehicle advanced systems, by designing ADAS that are more suitable to the drivers. As a practical example: the Lane Departure Warning (LDW) warns the driver when the left/right lane is crossed without using the indicator. However, blinkers are used only half the time before a lane change [18] and, therefore, the LDW might warn the driver in situations in which s/he is in full control of the vehicle (for example, during an overtaking without blinker activated), causing a nuisance to the driver. If this situation occurs frequently, the driver might get so annoyed by the system that might deactivate the LDW, eliminating the possible safety benefit brought by the system. If the human behaviour could be modelled more precisely, it would be possible to discriminate between an intentional lane crossing and (simply) an unintended lane crossing (with the LDW warning the driver only in the second case). Then, driver acceptance of the LDW could be increased. Similar examples could be found for other ADAS such as Forward Collision Warning (FCW) and Blind Spot System (BDS). Then, the driver intention detection module might be used jointly with other systems to warn the driver about risky behaviours or might be used for the communication with other ADAS. For instance, the lane change detection module could be implemented with a surrounding vision system or with a blind spot information system to prevent the driver from a dangerous overtaking manoeuvre (if an oncoming vehicle is spotted and, at the same time, a lane change intention is detected).

The Driver Intention Detection Module developed within the DESERVE project aims at modelling and predicting the driver's behavior at the tactical and operational levels of the Michon's model [20]. Among the maneuvers taken into consideration for the prediction of driver's intent, the most researched are the lane change, the turning left/right, the braking and the lane keeping. For the scope of the DESERVE project, the focus will be placed on the prediction of lane changes (and possibly of overtaking) with the final aim of improving the acceptance of ADAS. If a reliable lane change intention was developed, the warning could be issued only when needed: ADAS designed in such a way could increase driver's acceptance and reach a higher benefit with respect to road safety.

In the field of lane change intention detection, several researches have been already performed. One of the main authors on this topic is Salvucci. He applied the model tracing technique associated to a computational driver model to detect driver's intention to change lane [29]. Model tracing techniques were originally used for intelligent tutoring to predict students' possible next steps in problem solving. In the study of [29], data from the vehicle (steering wheel angle, accelerator depression, lateral position, longitudinal distance and time headway to a lead vehicle, longitudinal distance front and back, to vehicles in adjacent lanes) and from the environment (presence or absence of a lane to the left and right of the current travel lane) were used to build the model. Based on the information, the model calculates a desired steering angle and the accelerator position. The model performed well when tested both at the driving simulator and in the real vehicle, reaching a reliable detection of the maneuver after 1 second.

In a later research work [6], the authors developed and implemented a real-time lane change intent detection system which could go beyond the traditional offline implementation. The authors made use of information collected from the vehicle (steering wheel angle, yaw rate and blinker state signal), the Adaptive Cruise Control (distance to the lead vehicle, the relative speed, time gap to the vehicle in front and the difference between the current speed and the desired speed), the Lane Departure Warning (vehicle lateral deviation, lane curvature and vehicle yaw angle), the Side Warning Assist (occupancy and speed state within a critical zone) and the head position (head motion, head yaw and head pitch), adopting a time window of 2 second to trace the past events. A classifier based on relevance vector machines (RVM) was used for the lane change intent. The results show that, for a good prediction of the lane change intention, the inputs from the direct observation of the driver (head-viewing camera) are relevant and that the quality of the classification is improved (unreliable detections are beyond 3 seconds). In a later article [15], a multiclass Support Vector Machine (SVM) algorithm associated to a Bayesian filtering approach to predict lane change intention was used. The variables used as inputs for the algorithm were the lateral position of the vehicle (obtained from a lane tracker system), the steering angle, the first derivative of the lane position and the first derivative of the steering angle. The research was formulated as a multiclass classification problem with three possible outcomes: left lane change, right lane change and no lane change. On top of the multiclass classifier, a Bayesian Filter (BF) in order to improve the reliability of the predictions was used. The comparison between the SVM algorithm alone and the combination of SVM and BF shows that, in the first case, many false alarms were observed but the precision was increased by adding the Bayesian Filter, reducing average prediction times. Most of the lane changes are predicted almost 1.3 seconds before the lane crossing with a maximum prediction horizon reaching 3.3 seconds. The authors reported that further improvements might be brought by inclusion of other variables as the distance to the vehicle in front and the speed difference with the vehicle in front.

Overall, despite the knowledge acquired concerning the prediction of driver's intention to start a lane change, the topic is still interesting because the problem of lane change intention has shown to be extremely challenging. In particular, for having a more reliable prediction of driver intent, three aspects should be considered:

- to increase the precision of the prediction algorithms;
- to augment the detection time prior to the lane change;
- to decrease the number of variables to predict the lane change (not all the sensors used in the previous studies are available in common vehicles).

In addition, as pointed out by previous research [7], there are aspects which should be considered when designing a study to infer driver's intention prediction:

- type of inputs to be used: CAN data (steering wheel angle, pedal position, turn indicator), lane position sensor/camera (lateral lane position and standard deviation) and sensors for behavior data (head motion, eye motion foot and hands positions).

- type of algorithm to be adopted for the analysis: Support Vector Machines, Bayesian Nets, Hidden Markov Models
- material to be employed for the experiment: real vehicle (naturalistic or imposed) or driving simulator.

Regarding the first aspect, the results highly improve when measures of driver behavior are included, especially the head motion. However, this information is, usually, not available in common vehicles and, therefore, this feature should be further analyzed.

10.3 HMI Design Flow: The DESERVE Approach

In order to develop an HMI concept for ADAS capable of generating positive experiences during the driving task, a design workflow of 5 steps was used:

1. Collecting the state of art and last trends in the automotive HMI designing;
2. Defining three different HMI concepts;
3. Preliminary testing the three HMI concepts by a focus group;
4. Testing the best 2 concepts by a user test at driving simulator;
5. Defining the final concept.

The HMI was designed in order to allow adaptation strategies that takes into account the inputs provided by the driver model.

10.3.1 Different Approaches in the HMI of the Preventing Warning Systems: A State of Art in a Glance

From the point of view of the on-board human machine interface correlated to the different type of preventive accident systems, the evolution of HMI for ADAS could be clustered in three main phases.

It is possible to name the first era of preventive accident systems HMI as *warning era*. Most of the active and preventive systems above mentioned, which are not expected to be automatically actuated, are at the end a kind of *warning based systems* as they are aimed at increasing the driver awareness thanks to the support of technologies. The corresponding HMI is therefore based on alerts and aimed at delivering to the drivers immediately potential risks so to restore a safe situation for the driver.

The second phase coincides with an important transformation induced by the active and preventing safety systems evolution moving from being only activated by on-board sensors to a larger spectrum of sensors including both vehicle, other vehicles (Vehicles to Vehicles – V2V) and infrastructure

(Vehicles to Infrastructure – V2I). This technological evolution is creating so-called *cooperative ADAS* perspective [16] where preventive capabilities of such systems is allowed by the connection of the infrastructure. In terms of HMI, it is evident that vehicles are not necessary and exclusively oriented towards a dimension characterized by warning-based interfaces. Although this mechanism tends to persist, as well as to be necessary, it is also evident that within a system characterized by a high level of cooperation, the warning-based system might be easily replaced by a *recommending-based mechanism*. In other words, if vehicles are able to mutually recognize each other, as well as to cooperate for exchanging information and data, the system for supporting the driver will be aimed at sharing behavioural choice among the cars, rather than imposing and reporting imminent dangers. D3COS EU project (www.d3cos.eu) – among its results – have firstly proposed such promising concept in HMI for preventive accident systems [29]. This new dimension represents a real shift of paradigm going towards an increasing level of automation.

The third phase is characterized by the integration between the cooperative and the warning-based dimensions from one side, and the increased level of automation in cars (according to SAE Standard J0316) from the other. In this situation, expected HMIs will raise even more complex issues. Firstly, if on one hand it is true that automation will set the driver free from the necessity of constantly driving the vehicle, on the other hand, the driver is obliged to continuously monitor the correct functioning of the whole system. In a pioneering work, [1] expressed the idea of a sort of *irony* hiding behind the concept of automation. In fact, if theoretically speaking, the purpose of automation is to exclude the user from the driving tasks, in practices autonomous systems tends to encourage even more the participation of the driver, who must continuously monitor the correct functioning of the mechanism. The more the vehicle is autonomous, the more the driver is responsible for the only monitoring and the design issues for HMI designers is how to provide the best monitoring and to re-allocate the control to the drivers in the most effective and quicker way.

10.4 HMI Concepts Design

The three HMI concepts developed within the DESERVE project included the information normally displayed in the dashboard (i.e. speedometer, odometer, fuel level and water temperature information, diagnostic telltales, etc.), ADAS information support (i.e. lane change assistance system, nigh view, parking aid, adaptive cruise control, etc. as well as drowsy driver alert system) and navigation information.

Moreover a particular attention was dedicated to the design layout of the drowsy driver alert system. Drowsiness detection can be used to give a direct warning to the driver (explicit drowsiness) or as an input for an HMI reconfiguration strategy (implicit drowsiness). These two different strategies for drowsiness management were applied to all the three HMI concepts, obtaining hence 6 concepts to test. For the explicit drowsiness a warning is delivered to the driver with an icon and a message. For the implicit drowsiness ADAS sensitivity is set to the highest level. Once the driver takes a break, the ADAS configuration s/he set before is restored.

The user interface deploys 17 functions: 13 of them are ADAS, 2 are Safety Assistance Systems, and 2 are IVIS (In-Vehicle Information System), as listed in the following:

1. Lane change assistance system (ADAS);
2. Night vision system with pedestrian detection (ADAS);
3. Rear view camera system (Safety Assistance);
4. Surround view (Safety Assistance);
5. Lane departure warning (ADAS);
6. Pedestrian safety system (ADAS);
7. Collision warning system (ADAS);
8. Emergency braking ahead (ADAS);
9. Rear approaching vehicle (ADAS);
10. Adaptive high beam assist (ADAS);
11. Adaptive cruise control (ADAS);
12. Curve warning system (ADAS);
13. Intelligent park assist (ADAS);
14. Traffic sign recognition (ADAS);
15. Driver impairment warning system (ADAS);
16. Navi/Map info (IVIS);
17. Setting menu (IVIS).

10.4.1 Concept 1: Holistic HMI

In the Holistic HMI concept all the HMI elements (I/O) are centralized in front of the driver. The Instrument Panel Cluster (IPC) is the main visual output channel, while the steering wheel (SW) is the main input channel.

The HMI elements are listed as follows: i) IPC display 12"; ii) SW commands; iii) Left stalk commands; iv) Buttons; v) Knobs.

The instrument panel cluster was divided in three areas. In the central area the following information are delivered: lane change assistance system, night vision system with pedestrian detection, rear view camera system, surround view and setting menu.

The left area is mainly dedicated to the hazard warnings: lane departure warning, pedestrian safety system, collision warning system, emergency braking ahead, rear approaching vehicle, adaptive high beam assist, adaptive cruise control, and curve warning system are displayed.

In the right area the following information are delivered: intelligent park assist, traffic sign recognition, driver impairment warning system and navigation.

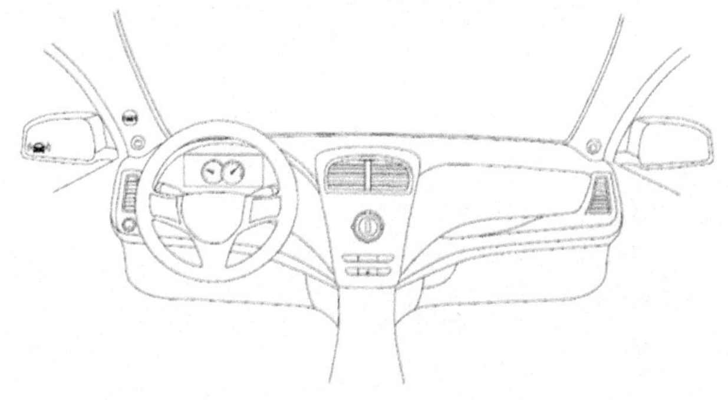

Figure 10.2 Holistic HMI concept, that shows: IPC display 12"; SW commands; left stalk commands; buttons; knobs.

Figure 10.3 Holistic HMI layout.

Figure 10.4 Holistic HMI layout with the user menu in the central area.

Figure 10.5 Holistic HMI layout with the lane change assist in the central area.

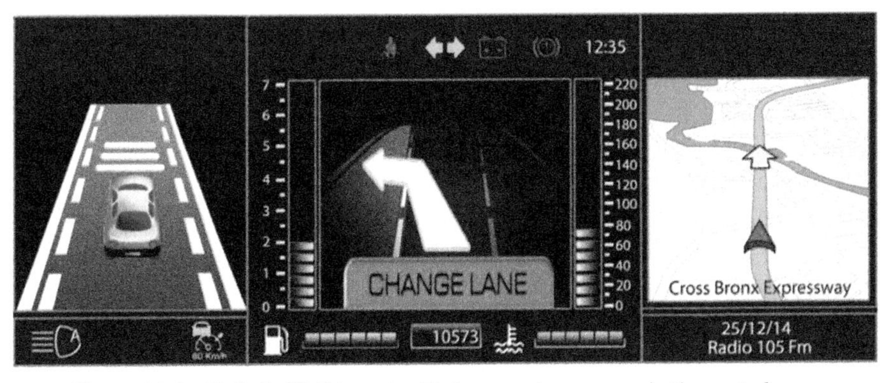

Figure 10.6 Holistic HMI layout with the rear view camera in the central area.

Figure 10.7 Holistic HMI layout with the night vision system in the central area.

Figure 10.8 (A-B-C-D) Holistic HMI left area with: lane departure warning, collision warning, Rear approaching vehicle system, pedestrian safety system.

10.4.2 Concept 2: Immersive HMI

The second concept is totally different from the previous one. While the Holistic HMI concept centralizes all the info and the interaction with the driver in front of him/her, the Immersive HMI concept distributes the interaction along the dashboard and the windscreen.

The HMI elements of concept 2 are listed as follows: i) 3,5" IPC display; ii) Touch Display 8,5" in the dashboard; iii) Head-up display for the windscreen; iv) SW commands; v) Left stalk commands; vi) Buttons; vii) Knobs.

In the concept 2 the area dedicated to the hazard warnings was moved in the middle of the instrument panel cluster, while the navigation, the rear view camera, the night vision system, radio/multimedia, phone and menu applications were moved to the dashboard display. The head-up display delivers traffic sign recognition and lane change assist information on the windscreen.

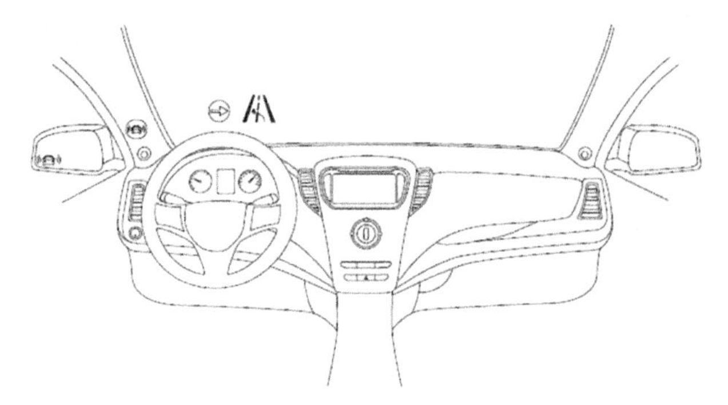

Figure 10.9 Immersive HMI concept shows: 3,5" IPC display; touch display 8,5" in the dashboard; head-up display for the windscreen; SW commands; left stalk commands; buttons; knobs.

10.4.3 Concept 3: Smart HMI

The third concept replaces the dashboard display with a nomadic device (ND – i.e. smartphone/tablet). The HMI can reconfigure itself according to ND size.

The IPC display has the same structure of that one of concept 2. The difference is that in the Smart HMI concept the 3,5" display of concept 2 was integrated by adding, for example, a 7" tablet (as in Figure 10.7) seamlessly connected with the car system. Drivers just connect the phone with a cable and immediately s/he gains access to ND applications using dashboard/steering-wheel buttons. The ND can provide also the access to further automotive applications. Driver can define what kind of information has to be shown in the ND: the ND is able to manage the infotainment functions and some ADAS applications.

Figure 10.10 Immersive HMI concept: instrument panel cluster display.

Figure 10.11 Immersive HMI concept: dashboard display.

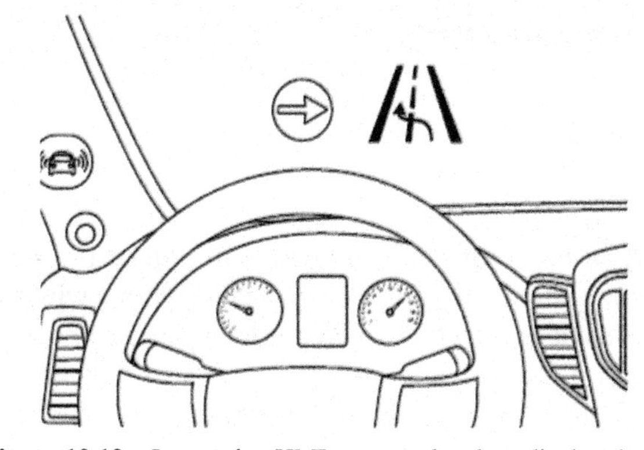

Figure 10.12 Immersive HMI concept: head-up display details.

The HMI elements of concept 3 are listed as follows: i) Display 3,5" in the IPC: ii) Touch Display of the nomadic device set into the dashboard; iii) SW commands; iv) Left stalk commands; v) Buttons; vi) Knobs.

10.5 Preliminary Testing by Focus Group

As Morgan described, "in essence, focus groups are special occasions devoted to gathering data on specific topics [21]". Using a focus group leads to evaluate preliminary concepts and in this case it is a useful technique to evaluate the proposals explained before [28], [35] having in mind that focus group is a technique deeply used in automotive field to evaluate user experience regarding HMI concepts [2, 9–12, 17].

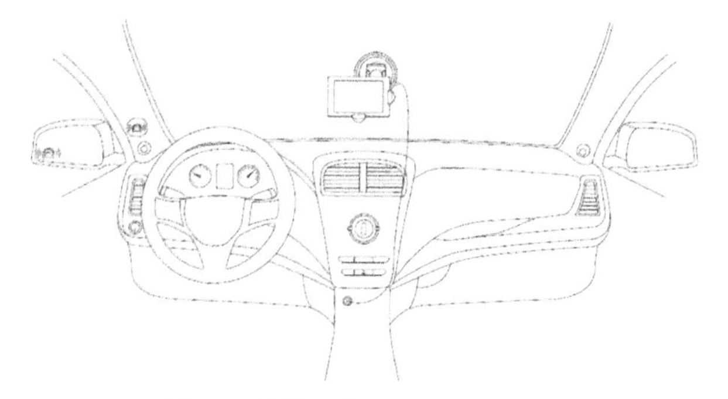

Figure 10.13 Smart HMI concept.

Figure 10.14 Smart HMI concept: Nomadic device with night vision system.

10.5.1 Participants

Sample is composed by 7 participants with a range of age between 25 and 39 years old (M = 31.71; SD = 4.06). Around 30% drive between 10000–15000 km/year and around 45% more than 20000 km/year. All drivers run at least once a day during the last year. 2 drivers run than 10% of their total driven by city, other 2 drive between 20–25%, and 3 run at least 40% or more of their driving in city. Moreover, around 40% drive usually on dual carriage way, and 30% run on highway and in similar percentage, 30%, drive on main roads.

10.5.2 Results

Participants discussed and exchanged points of views about HMI. They gave scores about degree of utility, easy to use, easy to learn, visual clarity, if the concepts were intuitive, degree of accessibility, and degree of driver annoyance and finally they provide a global value.

The HMI concept 1 (with explicit drowsiness) was considered as very useful, enough easy to use, with the most visual clarity and degree of accessibility among all the options presented. Having information located in same area is positive to avoid distraction and the three delimited areas for presenting information are pleasant. In general, alternative to concept 1 (with implicit drowsiness) is less appreciated than the original one. Scores are lowest than previous concept and the absence of drowsiness icon is missing by focus group participants.

Regarding HMI concept 2 (with explicit drowsiness), most of the participants appreciated to have information on HUD, moreover to have primary information in a different place from secondary one is a positive attribute. Besides, this concept seems to be a bit more easy to use and to learn and more intuitive. Concept 2 bis (with implicit drowsiness) is measured as intuitive and have visual clarity. Once more, HUD information is well appreciated by focus group participants. Anyway, it should be necessary to take into account that drivers are not being confident to manage drowsiness without a detailed icon.

Concept 3 (with explicit drowsiness) is not really appreciated from an aesthetically point of view. It is enough useful, easy to use and learn and enough intuitive. Focus group participants liked the possibility to place tablet where they prefer although it could mean less frontal vision. Last concept (n. 3 with implicit drowsiness) showed participants the least acceptable one. Although it will be positive to place the table according the wishes of drivers the general impression of having information in this way is not positive, even if it is having in mind that there is not drowsiness icon.

10.5.3 List of the Winning Features and Redesign Recommendations

As it can be observed in the radar chart which summarizes the HMI evaluation for the six concepts, concept 3 and its alternative, concept 3 bis (with implicit drowsiness) were the concept less valued. This concept "3 bis" is the concept which is considered more annoyed. Concept 2 had the highest average score for the global evaluation but concept 1 is closed to concept which adds information on a HUD and touch dashboard. Concept 1 stands out by its accessibility, utility and visual clarity and concept 2 is highlighted by its feature to be easy to use and it is a bit more intuitive.

During the session participants pointed several issues that should be taking into account:

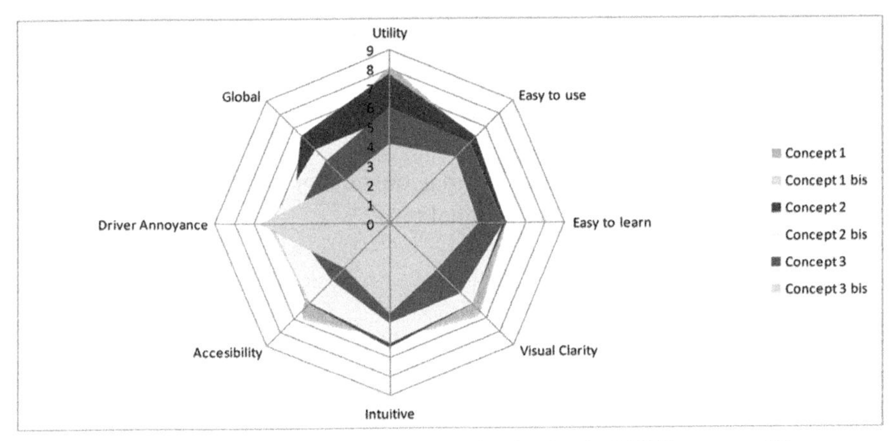

Figure 10.15 Radar chart summarizing HMI evaluation for the 6 HMI concepts. Bis concepts are concept 1, 2, 3 with implicit drowsiness.

- Summarizing the best option should have drowsiness icon.
- Option concept 1 and concept 2 are the best.
- The possibility to have HUD information is really appreciated.
- Participants suggested having in HUD the following information: traffic signals, gap for ACC, navigator system (with arrows and distances).
- For traffic signal information, it is very important to them to maintain this information available because sometimes you forgot this information (e.g. when you are running by a road and you forgot which was the speed limit).
- Information should be very clear and concise.
- It should be a great idea to have the possibility to select where you want to have the navigation system.

10.6 Users Test at Driving Simulator

As a final step for the definition of the overall HMI concept the two winning option from the focus group, namely concept 1 and concept 2 with the explicit drowsiness icon, were tested with users on a driving simulator in order to identify the final DESERVE HMI concept configuration. Each user was interviewed alone by a usability expert gathering comments and suggestions about the different ADAS function disposition and visualization.

Among the 13 ADAS functions developed for the DESERVE project, it was decided to test only 4 ADAS functions that were considered representative of the main HMI concept logic. In particular the following ADAS functions were widely tested with users:

1. Forward collision warning – with acoustic signal type 1.
2. Rear view camera system.
3. Lane change assistance system – with acoustic signal type 1.
4. Drowsiness icon – with acoustic signal type 2.

10.6.1 Participants

Sample is composed by 30 participants (20 Male and 10 female) with a range of age between 23 and 62 years old (M = 32.17; SD = 7.15). The majority of participants achieved a Master's degree.

The 30% of participants drive more than 20.000 km/year and the remaining between 10000 and 15000 km/year (M km/year = 15600; SD = 6931.18).

10.6.2 Procedure

After a brief explanation of test objective and some questions on personal data, user where asked to seat on the driving simulator and imagine to be inside their car, at the driving place with a dashboard of your car in front where some information about the car, its functioning and so on are displayed. Before assessing the solutions users where asked to practice a little with the driving simulator and to count the stars that appear on the road.

In particular user where asked to evaluate on a 7 point scale:

- The suitability of the HMI concept tested;
- The comprehensibility of the information displayed;
- The number of the information displayed;
- The pleasantness from a graphical point of view of the HMI concept tested;

10.6.3 Results

From the analysis of the different part of HMI concept test, concept 1 seems to be the preferred one even if the difference with the percentage of users that prefer concept 2 is not statistically significant. Despite this result the 60% of users would like to have the warning information in the central part of the display instead of in the lateral part. The functions representation seems quite clear for all users, only the adaptive light control and the adaptive cruise control icon should be re-designed. Considering the result of the task that

Figure 10.16 Proposed change to create the final DESERVE HMI concept.

Figure 10.17 Final DESERVE HMI concept: warning area.

Figure 10.18 Final DESERVE HMI concept: rear view camera.

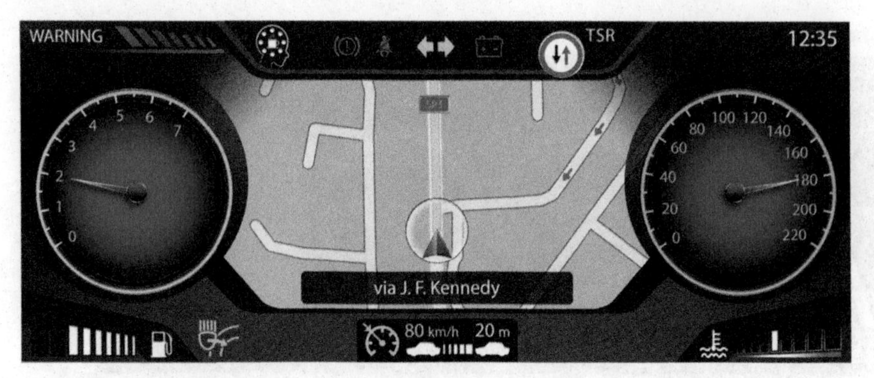

Figure 10.19 Final DESERVE HMI concept: navigation.

asked users to build their own solution, almost all distributed all functions in the same central display.

Thanks to users' feedbacks, the final DESERVE HMI concept has a single display with the warning functions in the central area and the gauges in the lateral part of the display.

10.7 Conclusions

Most cars today contain heterogeneous ADAS that support safe and clean driving. Because the pattern of factors in the automotive domain is constantly changing (new technologies and devices on board, new infrastructure, new mobility concepts, new trends in pollution prevention), the accident characteristics of the transport domain are also changing. As a consequence, also the research in that domain changed perspective, starting to investigate the human factor in order to improve safety and to prevent accidents. Even if it is not feasible to exactly predict the next accident, it is possible to anticipate some decisive characteristics of future accidents, as driver's misbehaviour. All these features concur in defining a new concept of ADAS system as a support and sometimes as a partner for drivers during task accomplishment and no more as a mere substitute.

Since nowadays more and more ADAS function are going to be implemented in current vehicles, the need for a unique Human Machine Interface is becoming an issue that reflects the increasing complexity of the entire system, whereby the driver has to deal with different devices and different interaction strategies. The aim of this work was in fact to identify the most suitable HMI concepts that allow an easy integration of different ADAS function in order to guarantee the safety of the introduction of any new element.

Acknowledgments

This study has been conducted thanks to the work and the experience of many people, besides the authors of this chapter.

We would like to express our thanks to Luana Baldassini for fruitful discussions, support, reviews and forward looking attitude.

References

[1] Bainbridge, L. (1983). Ironies of automation. *Automatica*, 19(6), 775–779.

[2] Barfield, W. & Dingus, T. (1997). Human factors in intelligent transportation systems Mahwah, NJ: Lawrence Erlbaum Associates.

[3] Barrera Murphy, N. & Knoblauch, R. (2004). *Hispanic Pedestrian and Bicycle Savety. Report of Focus Group Discussions in Washington, New Cork, Miami and Los Angeles*. http://safety.fhwa.dot.gov/ped_bike/docs/fhwanhtsa/fhwahtsa.pdf

[4] Bohlin, N. I. (1967). A statistical analysis of 28,000 accident cases with emphasis on occupant restraint value (No. 670925). SAE Technical Paper.

[5] Chowanetz, F., & Rigoll, G. (2011, October). A large-scale LED array to support anticipatory driving. In *Systems, Man, and Cybernetics (SMC), 2011 IEEE International Conference on* (pp. 2087–2092). IEEE.

[6] Doshi A., Morris, B., Trivedi M. (2011). On-road prediction of driver's intent with multimodal sensory cues. IEEE Automotive Pervasive Computing 10(3), 2011.

[7] Doshi A., Trivedi, M. M. (2011). Tactical driver behavior prediction and intent inference: A review. 14th International IEEE Conference on Intelligent Transportation Systems (ITSC), 2011. European Commissions, *Road safety statistics at regional level*, (2014), Source: Eurostat and DG Move. http://ec.europa.eu/eurostat/statistics-explained/index.php/_Road_safety_statistics_at_regional_level consulted on 14/12/2015.

[8] Green, P. & Brand, J. (1992). Future in-car information systems: input from focus groups (SAE paper 920614).

[9] Hof, T., Conde, L., Garcia, E., Iviglia, A., Jamson, S., Jopson, A., Lai, F., Merat, N., Nyberg, J., Rios, S., Sanchez, D., Scheineider, O. S., Seewald, P., Weerdt, C. V. D., Wijn, R. & Zlocki, A. (2012). D11.1: A state of the arte review and user's expectations. EcoDriver Project.

[10] Jeon, M., Schuett, J., Yim, J.-B., Raman, P. and Walker, B. N. (2011). ENGIN (Exploring Next Generation IN-vehicle INterfaces): Drawing

a new conceptual framework through iterative participatory processes. *Adjunct Proceedings of the 3rd International Conference on Automotive User Interfaces and Vehicular Applications (AutomotiveUI'11).*

[11] Kaufmann, C. Pereira, M., Simoes, A., Lancelle, V., Bruyas, M. P., Britschgi, V., Diez, J. L., Garcia Quinteiro, E. & Turetschek, C. (2010). A focus group approach towards an understanding of drivers' interaction with in-vehicle technologies. In J. F. Krems, T. Petzoldt & M. Henning (Eds.). *Proceedings of the European Conference on Human Interface Design for Intelligent Transport Systems*, Berlin, Germany, April 29–30 2010 (pp. 389–399). Lyon: Humanist Publications.

[12] Knobel, M., Schwarz, F., Palleis, H., & Schumann, J. Towards Designing Experiences for Advanced Driving Assistance Systems. In *Workshop User Experience in Cars. In conjunction with 13th IFIP TC13 Conference on Human-Computer Interaction (INTERACT 2011)* (pp. 05–09).

[13] Krueger, R. A. (1988). *El grupo de discusión. Guía práctica para la investigación aplicada.* Madrid: Ediciones Pirámide.

[14] Kumar P., Perrollaz M., Lefevre S., Laugier C. (2013). Learning-Based Approach for Online Lane Change Intention Prediction. IEEE Intelligent Vehicles Symposium, 2013.

[15] Laquai, F., Gusmini, C., Tonnis, M., Rigoll, G., & Klinker, G. (2013, October). A multi lane Car Following Model for cooperative ADAS. In *Intelligent Transportation Systems-(ITSC), 2013 16th International IEEE Conference on* (pp. 1579–1586). IEEE.

[16] Larsson, P., Esberg, I., van Noort, M., Willemsen, D., Garcia, E., Fahrenkrog, F., Zlocki, A., Scholliers, J., Koskinen, S., Várhelyi, A., Schönebeck, S. (2012). "Test and evaluation plans". Deliverable D7.4 of the InteractIVe project.

[17] Lee S. E., Olsen E. C. B., Wierwille W. W. (2004). A Comprehensive Examination of Naturalistic Lane-Changes. National Highway Traffic Safety Administration, DOT HS 809 702, 2004.

[18] Lie, A., & Tingvall, C. (2002). How do Euro NCAP results correlate with real-life injury risks? A paired comparison study of car-to-car crashes. *Traffic Injury Prevention*, 3(4), 288–293.

[19] Michon, J. A., A Critical View of Driver Behavior Models: What Do We Know, What Should We Do?, Human Behavior and Traffic Safety, pp. 485–524, 1985.

[20] Morgan, D. L, Krueger, R. A., King, J. A. & Scannell, A. U. (1998). *The Focus Group Kit, Volumes 1–6*. Thousand Oaks, CA: SAGE Publications.

[21] National Highway Traffic Savety Administration (NHTSA) (2008). *Summary of Focus Group Findings.* http://www.nhtsa.gov/staticfiles/DOT/NHTSA/Rulemaking/Rules/Associated%20files/5StarFocusGroup.pdf

[22] Olsheski, J. D., Walker, B. N., & McCloud, J. (2011, October). In-vehicle assistive technology (IVAT) for drivers who have survived a traumatic brain injury. In *The proceedings of the 13th international ACM SIGACCESS conference on Computers and accessibility* (pp. 257–258). ACM.

[23] Parasuraman, R., & Riley, V. (1997). Humans and automation: Use, misuse, disuse, abuse. *Human Factors: The Journal of the Human Factors and Ergonomics Society*, 39(2), 230–253.

[24] Peltzman, S. (1975). The effects of automobile safety regulation. *The Journal of Political Economy*, 677–725.

[25] Pinotti, D., Tango, F., Losi, M. G., & Beltrami, M. (2014). A model for an innovative Lane Change Assistant HMI. In *Proceedings of the Human Factors and Ergonomics Society Europe Chapter 2013 Annual Conference.*

[26] Rasmussen, J., Skills, Rules and Knowledge; Signals, Signs, and Symbols, and Other Distinctions in Human Performance Models, IEEE Transactions on Systems, Man, and Cybernetics 13 (3), pp. 257–266, 1983.

[27] Rubin, J. & Chisnell, D. (2008). Handbook of Usability Testing. How to plan, Design, and Conduct Effective Tests. 2nd Edition. IN: Wiley Publishing.

[28] Salvucci D. D., Mandalia H. M., Kuge N., Yamamura T. Lane-Change Detection Using a Computational Driver Model. Human Factors 49(3) 2007, pp.532–542.

[29] Schmidt, A., Dey, A. K., Kun, A. L., & Spiessl, W. (2010, April). Automotive user interfaces: human computer interaction in the car. In *CHI'10 Extended Abstracts on Human Factors in Computing Systems* (pp. 3177–3180). ACM.

[30] Schmittner, C., Gruber, T., Puschner, P., & Schoitsch, E. (2014). Security application of failure mode and effect analysis (FMEA). In *Computer Safety, Reliability, and Security* (pp. 310–325). Springer International Publishing.

[31] Seay, A., Zaloshnja, E., Miller, T., Romano, E., Luchter, S., & Spicer, R. (2002). *The economic impact of motor vehicle crashes, 2000* (No. HS-809 446,). Washington, DC: US Department of Transportation, National Highway Traffic Safety Administration.

[32] Sharken Simon, J. (1999) [On line]. *How to conduct a Focus Group*. Recovered on August 18th, 2008 from: http://tgi.com/magazine/HowToConductAFocusGroup.pdf

[33] Treat, J. R., Tumbas, N. S., McDonald, S. T., Shinar, D., & Hume, R. D. (1979). *TRI-LEVEL STUDY OF THE CAUSES OF TRAFFIC ACCIDENTS. EXECUTIVE SUMMARY* (No. DOTHS034353579TAC (5) Final Rpt).

[34] Tullis, T. & Albert, W. (2013). Measuring the user experience. 2nd Edition. Morgan Kaufmann.

11

Vehicle Hardware-In-the-Loop System for ADAS Virtual Testing

Romain Rossi, Clément Galko, Hariharan Narasimman and Xavier Savatier

Univ. Rouen, UNIROUEN, ESIGELEC, IRSEEM 76000 Rouen, France

11.1 Introduction

Testing vehicular functions can be a very tedious task. The classical approach tries to tackle this problem using a multiple-stage validation and testing process. The first step is a Model-In-the-Loop (MIL) approach which allows quick algorithmic development without involving dedicated hardware. Usually, this level of development involves high-level abstraction software frameworks running on general-purpose computers. The second step is a Software-In-the-Loop (SIL) validation, where the actual implementation of the developed model will be evaluated on general-purpose hardware. This step requires a complete software implementation very close to the final one. The last step of this validation process is Hardware-In-the-Loop (HIL) which involves the final hardware, running the final software with input and output connected to a simulator. This proven process is very widely used in the transportation industry and has enabled the development of very high quality components which are then integrated into bigger systems or vehicles. Modern vehicles however integrate so many such components that the integration phase has become more complex and also requires a multi-step validation process. The final integration tests are performed on tracks or roads. While mandatory, these real-condition tests are limited because of multiple factors and have a very high cost.

Testing a complex system like a modern vehicle on a test track or on a real road involves complex and costly engineering. First of all, to be testable the vehicle must be fully or nearly-fully functional. This limits the testing opportunity to a very late stage in the development process and implies high engineering costs. Moreover, because the real-condition test is very constrained in time and space, the test coverage is not complete and only a very small variety of real-world conditions can be tested.

To address these limitations and lower the cost, modern ADAS (Advanced Driver Assistance Systems) development frameworks uses a virtual test bench approach where realistic simulator software and hardware are used to enable faster and less expensive tests with better coverage on complete vehicles. In this document, we propose a virtual testing system built on a chassis dynamometer which enables a complex test scenario to be applied early in the ADAS development process.

Our proposed system, named SERBER (Simulateur d' Environnement Routier integré à un Banc de test véhicule pour l'Evaluation de stratégies de gestion de l'éneRgie embarquée) aims to ease ADAS prototypes testing and at the same time, analyze the energy efficiency of the prototype system using the standard equipment of the chassis dynamometer. A previous version of this system has been published in [3], which presented the SERBER system and showed preliminary results.

11.2 State of the Art

In the automotive industry, car manufacturers use different ways to test and validate ADAS and other embedded systems. An extensive study of the state of the art in ADAS testing and validation methods can be found in [1]. These test methods can be grouped in two categories: test-bench tests and in-vehicle tests.

For test-bench tests, three approaches are usually used during the development cycle: Model-In-the-Loop (MIL), Software-In-the-Loop (SIL) and Hardware-In-the-Loop (HIL). In MIL, a model of the developed system is integrated in a simulation loop with models of vehicle dynamics, sensor, actuators and traffic environment. After successful MIL validation, the SIL approach allows to replace the tested model with a real software implementation for real-time operation validation. The last step, HIL, consists of a combination of simulated and real components in order to validate the functionality of the developed system on both hardware and software aspects.

Test-bench tests are very useful as they provide a safe, repeatable and reliable way to validate these embedded systems under a variety of operating conditions. This kind of tests also has some drawbacks. For example, the interaction with other ADAS is difficult to test as well as the integration in the vehicle system. A sample of a HIL test bench for complex ADAS is available in [2].

The second category of tests methods are in-vehicle tests. These tests require a prototype to integrate the developed system. Again, three approaches are commonly used: test-drives on test-tracks, test-drives on open-roads and Vehicle-Hardware-In-the-Loop (VeHIL).

The first two approaches are very similar and assume the prototype to be driven in real-conditions. The test-track allows control of some environment parameters (traffic, some weather conditions, road signs, road type and so on) but requires big infrastructures. The open-road tests require less dedicated infrastructures but are of limited use because of the difficulty to reproduce the needed conditions, and the underlying safety problems. Both of these methods are costly and time-consuming and can't be used early in the development cycle because they require heavy engineering efforts to have a fully functional prototype to drive.

A very interesting solution which combines nearly all the advantages of the previous methods without most of their drawbacks is the VeHIL approach. This kind of tests is a combination of the HIL and test-drives approaches. Functional as well as integration tests can be done easily and early in the development cycle. As the vehicle is physically locked on the chassis-dynamometer, this system greatly improves the safety of the tests. Because it is an indoor test, every environmental parameter (humidity, ambient light, temperature and so on) can easily be controlled and thus the repeatability of the test is ensured.

Existing VeHIL systems like the one described in [1] and currently used by [2] relies on mobile platforms (called Mobile Bases) to move targets (fake cars and pedestrians) in front of the tested vehicle in order to trigger the various embedded functions (pedestrian detection, ACC, AEB and so on). This setup however needs heavy infrastructure: the chassis-dynamometer is installed in a very large room (200×40 m) and the targets are moved at high speed by the Mobile Bases which can be dangerous for both the tested vehicle and the persons involved. Thus, the tests are remotely executed from a control room and the test area has to be evacuated.

11.3 Proposed System

To address the problems of existing VeHIL systems (large infrastructures, fast moving targets, hazard for people), we propose a system which associates a chassis dynamometer with multi-sensor road environment simulation software. The simulator uses a description of the virtual environment and the position of the vehicle to generate multi-sensors data. These data are then fed into the sensors of the real car placed on the chassis dynamometer. On the other way, motion data (speed, acceleration) are gathered from the chassis-dynamometer and used to update the simulated vehicle speed and position.

Our system, as seen in Figure 11.1 is mainly composed of three parts: the chassis-dynamometer, multi-sensor simulation software running in a computer and devices to feed the vehicle sensors like LCD screen and the CAN bus interface with synthetic data. The chassis-dynamometer is standard equipment, the main requirement is to be able to connect it with the simulation computer in order to read the vehicle actual speed and control the simulated slope by adjusting the friction force applied by the dynamometer. The simulation software is at the core of our system and is responsible for the generation of sensor data to be fed into ADAS sensors. We use the Pro-Sivic software dedicated to this kind of application. An introduction to Pro-Sivic can be found in [4]. The difficulties in our proposed system lies mainly in the way to fool

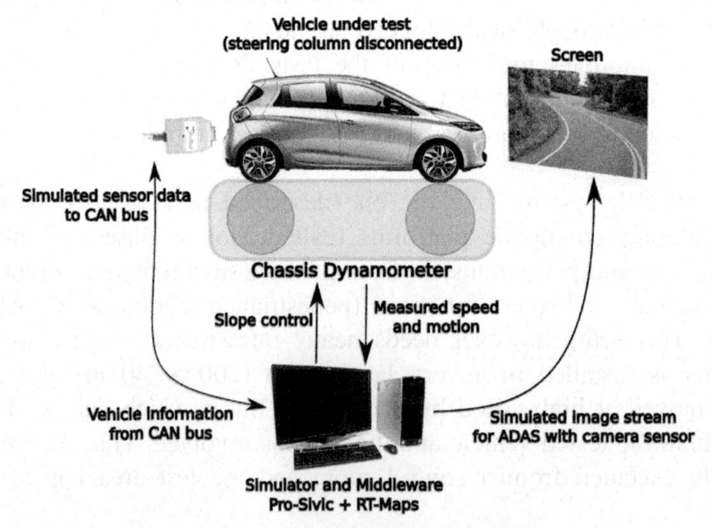

Figure 11.1 Overview of the SERBER VeHIL system.

(i.e. feed synthetic data into) the vehicle's sensors with the data produced by the simulation software.

Three ways can be used to fool sensors. The first way (full simulation) is to disconnect completely the ADAS and replace it with an electronic probe controlled by our system, which simulates the ADAS behavior completely. The simulated data (ADAS outputs) are sent directly into the vehicle internal communication bus (CAN) to be used by the other vehicular functions. The second way (sensor simulation) is to disconnect only the sensor part of the ADAS and replace it with an electronic probe. The simulator generates data according to the specification of the simulated sensor. The "signal processing" part of the ADAS is kept in the loop, so it can be tested by the system. This approach however requires the sensor to be separated from the main ADAS unit. The last way (stimulation) is to keep the full ADAS in the loop and send physical stimuli to the ADAS sensor through dedicated hardware. For example, an LCD screen can be placed in front of an embedded camera or a Hyper-Frequency generator can send signals to an embedded RADAR sensor.

This last solution is the preferred one, as it keeps the whole ADAS in the testing loop and limits the modifications done to the vehicle. So the objective of our work is to be able to simulate and fool every vehicle sensors. This approach however is very difficult to achieve for some kind of sensors, like inertial sensors and environmental sensors, or needs very complex stimulation hardware for RADAR and LIDAR.

With such a hardware-in-the-loop system, multiple scenarios can be implemented and tested in the safety and convenience of an indoor workshop. This system can be used for new ADAS prototyping as it is very easy to produce test-cases for the specific system under development. It can also be used to test the integration of multiple ADAS in a car, using a set of predefined test-cases to validate their interaction. It can also be used for very complex ADAS or fully-automated vehicle development where the embedded system relies simultaneously on multiple sensors to operate, because it is able to simulate nearly every aspect of the road environment at the same time.

Moreover, the use of a chassis dynamometer allows a simultaneous analysis of various performance indicators of the vehicle, including energy consumption and pollution. This coupling is a real benefit compared to traditional test setups and enables the early evaluation of the energy consumption impact of various changes in the ADAS systems. For example, the fuel consumption and pollution of a car equipped with Adaptive Cruise Control (ACC) can be continuously monitored as various ACC algorithms are developed and tested.

11.4 Hardware Implementation

The proposed system is implemented at IRSEEM facilities. A chassis-dynamometer is available in one of our technological platform and is used as a building block. This chassis dynamometer is a Horiba Vulcan 4WD with two independent axles. It provides real-time velocity information based on the real vehicle wheel speed. It can also apply a friction force equivalent to a 5% slope of the road. The control system allows interfacing through analog inputs and outputs, these are used to control the friction force to simulate the slope and to read the actual speed of the vehicle in real-time. A complete description of the used chassis-dynamometer is available from the Horiba website [5]. We use an analog input/output device from National Instruments to link the simulation computer with the chassis-dynamometer control system.

One of the main challenges in our proposal is the ability to generate synthetic data and feed the ADAS sensors with this data. The synthetic data generation for vision-based, RADAR-based and LIDAR-based sensors is handled by Pro-Sivic. The main problem is how to correctly stimulate the sensors to feed it with this simulation data.

11.4.1 Sensors Stimulation Solutions

Feeding sensors with simulated data is a key function of our system and a complex challenge. A vehicle can embed numerous sensors like cameras, inertial sensors, temperature sensors, rain sensors, odometer, LIDAR, RADAR, GPS and more. Because of the broad variety of sensor types, different approaches are needed to be able to control what the sensor reports to the ADAS processor.

For camera-based sensors, we use direct stimulation using a standard computer display placed in front of the camera. A first successful test was done with a 32" LCD screen. A display system using a projector would allow a bigger image surface and is currently being tested. Image field-of-view and distortions have to be taken into account for an accurate stimulation of the ADAS sensor. Special care must be taken in order to completely cover the sensor's field of view. This is especially difficult with wide-angle or fish-eye cameras and would require special setup.

The rain sensor can easily be triggered using a localized water diffusion device (sprinkler) actuated by a solenoid valve. This system can also be used to generate rain-like perturbation on camera sensors by directly applying water on the windshield. However, such solutions can produce perturbations which are not reproducible.

GPS simulation devices already exist for factory tests and are able to generate a controlled fake position to be interpreted by GPS receivers nearby. This kind of device could easily be integrated with our system to provide real-time positioning to the vehicle and embedded ADAS using GPS as a source of information. These systems are however costly and a direct transmission of generated NMEA frames to the ADAS is vastly more cost effective, but needs a small modification of the vehicle under test.

Likewise, real-time target generators for various types of RADAR (24 and 76 GHz) are available as off-the-shelve component. These systems cans also be coupled with the real-time simulation software to report the position and speed of simulated actors to the vehicle.

A recent paper [6] shows a possible implementation of a target simulator for LIDAR sensors. In this paper, a pulse generator is synchronized with the LIDAR in order to inject false object echo. Fully functional real-time target simulators are however to be demonstrated. This setup could be used in our system like the RADAR target simulator described above.

Inertial sensors are not covered yet. These sensors are usually deeply embedded in ECU and can be difficult to physically disconnect. An option is to physically move the vehicle using external actuators, but this implies heavy equipment. Another option is to open the ECU and physically replace the sensor with an electronic probe, which is time-consuming and difficult to achieve without complete documentation.

For Ultra-Sonic range-finder, two main solutions are possible. The first one is to use a sound generator simulating echo. The other one is to use small mobile targets located directly in front of the sensors. As these sensors are usually used only for low-speed maneuver and short-range detection, these mobile systems would not require a big infrastructure and can safely be used even in the presence of people.

Recently, Vehicle-to-Vehicle (V2V) and Vehicle-to-Infrastructure (V2I) communication has widespread with the 802.11p standard. These systems are used as a kind of virtual sensor providing the position, relative speed and status of the vehicles in the vicinity. Because of their operation, such systems are easy to connect with a computer. In our system we used one 802.11p modem to generate synthetic CAM and DENM messages to be interpreted by the vehicle under test.

Feeding the sensor with simulated data is not an easy task and each sensor has to be addressed differently. We plan to use sensor stimulation whenever possible, and fall back to sensor simulation and sending data in the CAN bus when stimulation is not feasible. Some kind of sensors appears to be relatively easy to stimulate (like cameras), others needs very complex and costly equipment (GPS, RADAR and inertial sensors).

For all these sensors, an alternative approach would be to have a cooperative software embedded in the ECU which would allow to overwrite actual measurements through the CAN bus (or another communication medium). While this solution seems unlikely to be possible on production vehicles; prototypes and test vehicles can be equipped with such debugging software, enabling a controlled and effective way to bypass sensors and feed synthetic data straight to the embedded processors.

11.4.2 Software Implementation

Our software runs on a high-end laptop computer and is based on two main building blocks: multi-sensor simulation software (Pro-Sivic) and a real-time middleware (RTMAPS). A block diagram of the complete system is presented in Figure 11.2.

To run the simulations, we used Pro-Sivic from Civitec. This real-time multi-sensor simulation software is a fusion between a driving simulator and a multi-sensor simulator. Pro-Sivic provides kinetic data and sensor data from

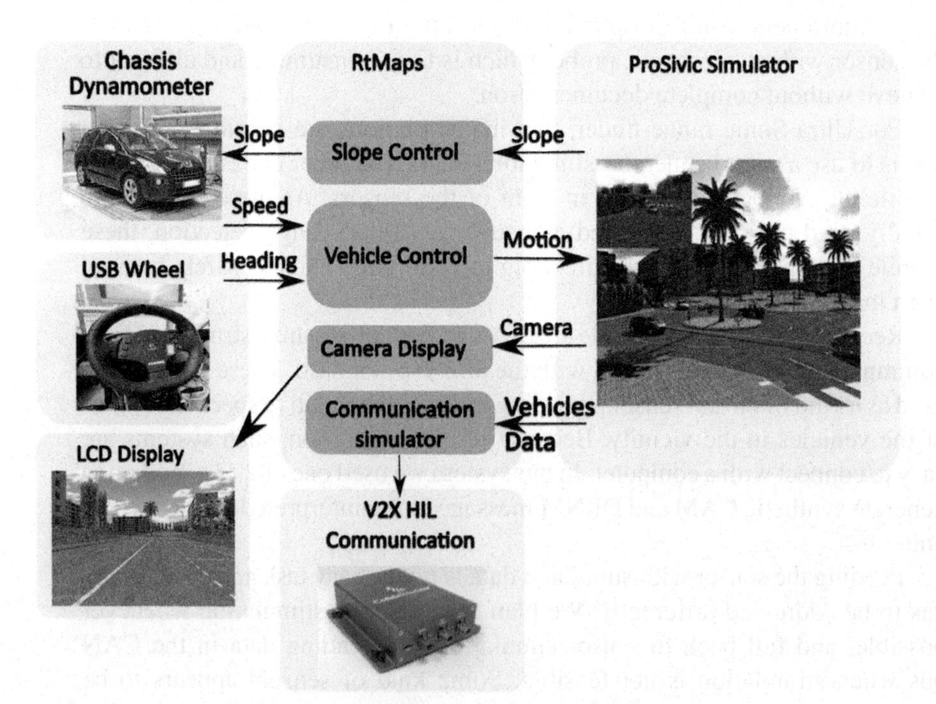

Figure 11.2 Block diagram of the SERBER system.

the simulated vehicle and can also be used as a driving simulator. A complete description of Pro-Sivic is given in [4]. Pro-Sivic is able to generate realistic video output which can be directly used to stimulate camera-based ADAS. A sample view of Pro-Sivic video output can be seen in Figure 11.3.

The other main software building block, RTMAPS from Intempora is a middleware which interconnects all other parts of the system. It is also used to produce CAN messages to be sent on the vehicle bus and perform other implementation-specific operations. RTMAPS is a component-based graphical programming framework to easily build multi-tasks or distributed applications. This software is described in detail in [7] and in this book (Part 1, Chapter 4). RTMAPS provides native interfaces to multiple simulation software, including Pro-Sivic, and also numerous components for device support (CANpeak, serial GPS, National Instruments I/O device and so on).

The most significant part of the RTMAPS diagram is presented in Figure 11.4. This diagram main task is to handle the communication between Pro-Sivic and the chassis-dynamometer; and to generate Vehicle-to-Vehicle and Vehicle-to-Infrastructure communication messages based on the simulation data.

Figure 11.3 Sample video output of Pro-Sivic.

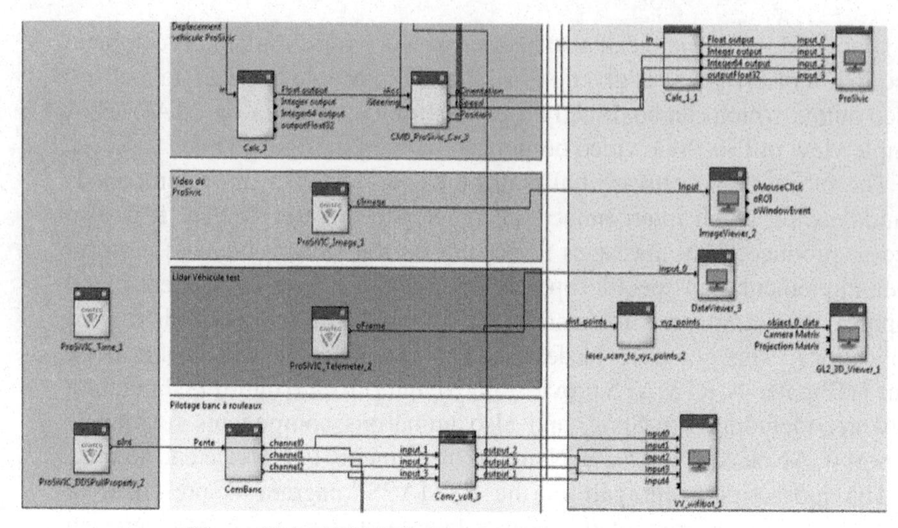

Figure 11.4 RTMAPS diagram of the system (extract).

Our chassis-dynamometer has a hardware limitation: the vehicle front wheels cannot turn when the vehicle is moving; or damage can occur. In order to prevent this, the test vehicles driving wheel is physically removed and replaced by a USB joystick connected to the computer. This allows lateral control of the virtual vehicle by the driver, in a way very similar to driving simulators, while the physical wheels stays in line with the chassis dynamometer.

11.5 Experimental Setup

In order to test our system, we equipped a small fully-electrical vehicle with an after-market ADAS system: a Mobileye 560. This ADAS, designed to be installed on the windshield, is based on a forward-looking camera and an integrated processor which performs real-time image processing. The main unit contains the camera and a processing device, and a separate display is used to inform the driver of the working state of the system and to show warnings. A Bluetooth connection allows using a dedicated application on a smartphone or tablet to display various data in addition to the one already shown on the small display. A picture of the system is shown in Figure 11.5 where the main unit is shown on the right, the small display in the middle, and a smart-phone running the dedicated application on the left.

The Mobileye system is able to detect and track many objects: pedestrians, other vehicles, speed-limit signs, and white lines. The position of the tracked

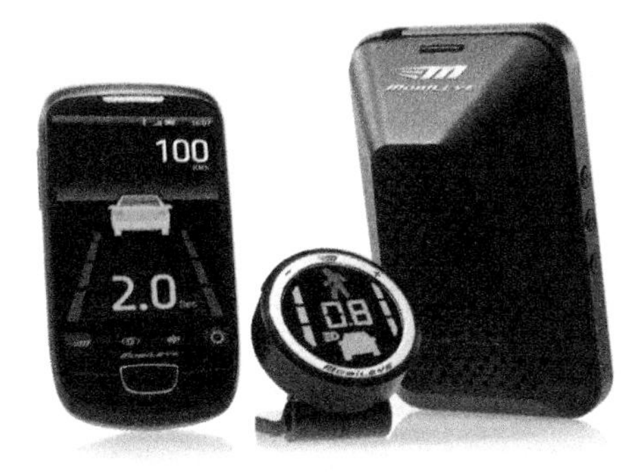

Figure 11.5 Mobileye 560 aftermarket vision-based ADAS.

objects, as well as the vehicle speed information gathered from the CAN-bus is used to detect dangerous situations and to warn the driver: risk of pedestrian collision, risk of forward collision, lane departure, over-speed and so on. All the processing is done inside the Mobileye main unit and only high-level information is available through a small display. The Mobileye system is described in details in [8] and up-to-date information is available in [9].

The V2V communication test bench is composed of two Khoda Wireless MK 2802.11p modems equipped with MobileMark SMW-303 multiband antennas. One of the modem is used to send CAM and DENM messages generated from virtual vehicles data. The other modem is used as an embedded unit in the vehicle under test. The data received by this second modem are used to update a dashboard HMI. An extract of the RTMAPS diagram responsible for the communication task is shown Figure 11.6.

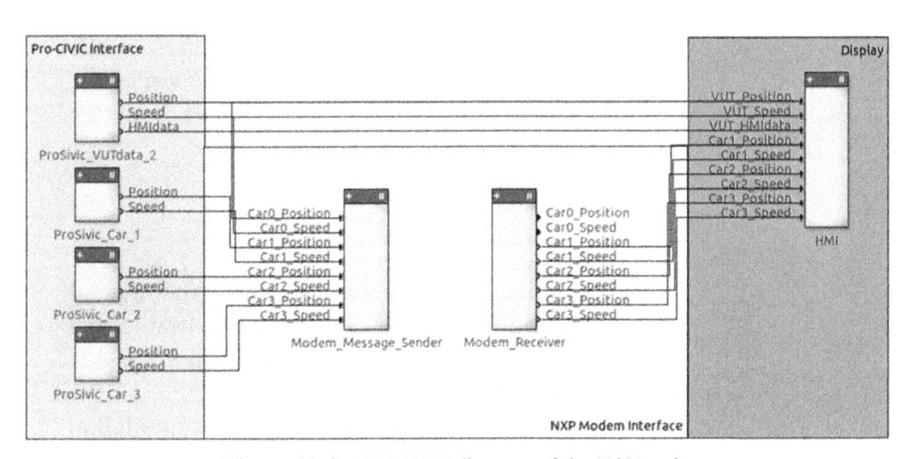

Figure 11.6 RTMAPS diagram of the V2V task.

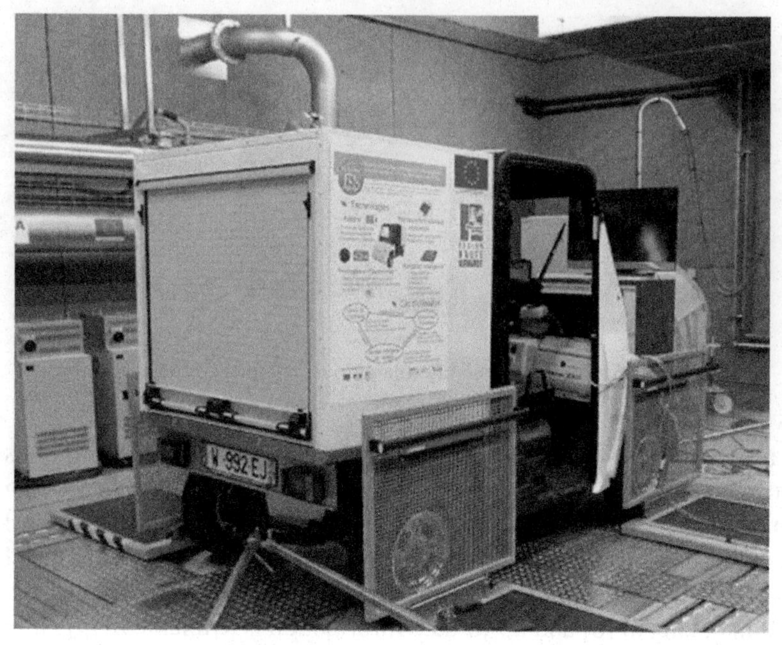

Figure 11.7 The Biocar test vehicle on the Horiba chassis dynamometer.

The test vehicle equipped with the Mobileye is placed on the chassis dynamometer, and an LCD screen is placed in front of the windshield, in the sight of both the driver and the Mobileye system. The Figure 11.7 shows a view of the test vehicle installed on the chassis-dynamometer. The LCD screen can be seen in front of the car.

The whole system was tested with an urban scenario and environment. This scenario is composed of a few roads with some buildings and trees; the traffic is simulated with four cars following a predefined path. The virtual car can freely move inside this environment and is directed by actions from the driver. A view of the urban scenario is shown in Figure 11.8.

11.6 Results

A first series of results have been obtained with the described experimental setup. The virtual car forward motion is completely controlled by the real vehicle controls (accelerator and brake pedals), while the lateral control is obtained from the USB driving wheel connected to the computer.

First, the integration of the chassis-dynamometer with the simulation software was tested. The real car speed is read and used to update the virtual

Figure 11.8 Overview of the urban environment in Pro-Sivic.

vehicle motion. In Pro-Sivic, the road slope under the vehicle is processed and this information is used to control the resistive torque applied by the chassis-dynamometer on the real vehicle. During the tests, the car driver can feel the resistive torque applied by the system on the vehicle wheels when climbing a slope, and has a feeling of free wheels when going down. The Figure 11.9 shows a picture taken near the driver's seat. Driving the car is natural and intuitive, just as if the car would be on a real road. The driving simulator use-case is not the main goal of this system but this first test proves the interest of the SERBER system even for ADAS which involves driver interaction.

The ADAS sensor stimulation abilities of the system were tested using the Mobileye. This test showed promising results as the Mobileye was fooled by the simulation and worked as if the car would be running on a real road. The lane departure warning and forward collision warning have been triggered by the corresponding simulated situations. The Figure 11.10 shows the lane departure warning being triggered when the car is crossing the road central line with the blinkers off. In this picture, the road is clearly seen on the LCD screen in the top right part. In the bottom left part, a tablet running the Mobileye application shows a graphical representation of the warning being triggered.

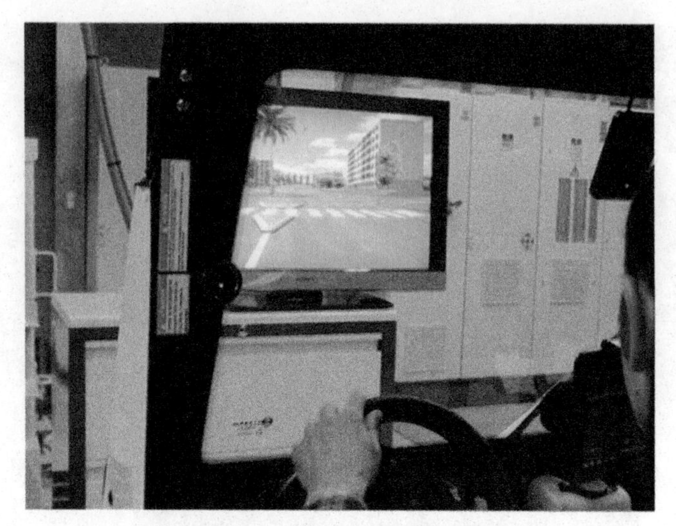

Figure 11.9 Inner view of the vehicle.

Figure 11.10 Lane departure warning triggered.

The last tested functionality is the V2V communication simulation. In this test, four virtual vehicles are simulated and their global positions are broadcasted by the 802.11p modem using CAM messages. Various DENM messages are also broadcasted by virtual vehicles. Another modem is embedded in the vehicle under test and receives these messages. An HMI is used to display this

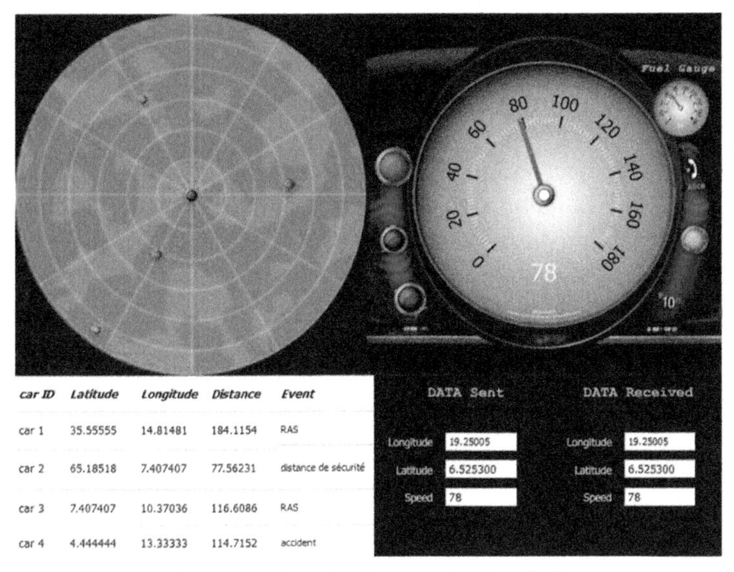

Figure 11.11 V2V Communication HMI.

data to the driver, using a RADAR-like circular representation. A snapshot of the HMI is shown in Figure 11.11.

11.7 Conclusion and Future Work

In this chapter, we have presented SERBER, our Vehicle-Hardware-In-the-Loop system which uses a chassis-dynamometer and a multi-sensor simulation software to create a kind of virtual reality platform for intelligent vehicles equipped with ADAS using sensors to gather information from the surrounding environment. The combination of the simulation software and the chassis-dynamometer allows applying the resulting force from a simulated slope to the real vehicle, while the sensor data generated by the simulation software are fed into the ADAS.

We discussed different way show the system can feed simulated data to sensors, both at the communication-bus level (CAN messages) and at the physical-stimuli level.

We described our current implementation based on Pro-Sivic, RTMAPS and a Horiba chassis-dynamometer and presented the first results obtained by a complete test using a small electrical car equipped with an after-market camera-based ADAS and 802.11p modem. The result presented in this paper shows the ability to fool an ADAS system based on a forward-looking camera. Various functions of the ADAS are triggered when corresponding situations are simulated: forward collision warning and lane departure warning.

The DESERVE project aims to provide an environment for ADAS design, development and pre-validation. In this context, SERBER provides a virtual testing platform enabling early tests of newly designed ADAS with realistic scenarios and testing environments. This system can also be used to validate multiple ADAS interaction on the same vehicle and aims to be a complete test and validation system for fully-autonomous vehicles.

SERBER is more compact and simpler to use than other VeHIL systems which use mobile bases to move fake cars at high speed in order to simulate other vehicles motion. In fact, our system can easily be installed on a standard chassis-dynamometer, if it can be controlled by software, requiring only minor physical modification of the facility.

The work presented in this chapter is a first step towards a complete simulation system able to stimulate multiple sensors in the tested vehicle. Currently, only camera-based ADAS and V2X communication systems can be stimulated.

The first area of improvement for the current system is the simulation and stimulation of additional sensors. A RADAR virtual target generator is currently being developed in order to fool RADAR-based ADAS like Adaptive Cruise Control (ACC) and Automatic Emergency Braking (AEB). A LIDAR target generator and GPS simulator can be integrated to provide a quite complete setup able to test realistic scenarios.

A second area of improvement is in the simulation environment and scenario. To be able to test corner-cases and complex interaction of various ADAS functions, sophisticated scenarios involving various road environments, pedestrians, other vehicles and driver behavior have to be designed and implemented.

Acknowledgment

The DESERVE project (Development platform for Safe and Efficient dRiVE, is a project funded by ECSEL-JU (http://ecsel.eu) and is available at http://deserve-project.eu

The SERBER project (Simulateur d'Environnement Routier integré à un Banc de test véhicule pour l'Evaluation de stratégies de gestion de l'éneRgie embarquée) is funded by Institut Carnot ESP (http://www.carnot-esp.fr).

References

[1] C. Galko, R. Rossi and X. Savatier, "Vehicle-Hardware-In-The-Loop System for ADAS Prototyping and Validation," in *IEEE International Conference on Embedded Computer Systems: Architectures, Modeling, and Simulation (SAMOS XIV)*, SAMOS Island, Greece, 2014.

[2] O. Gietelink, J. Ploeg, B. De Schutter and M. Verhaegen, "Development of advanced driver assistance systems with vehicle hardware-in-the-loop simulations," *Vehicle System Dynamics,* Vol. 44, No. 7, pp. 569–590, 2006.

[3] K. Athanasas, C. Bonnet, H. Fritz, C. Scheidler, and G. Volk, "VALSE-validation of safety-related driver assistance systems," chez *Intelligent Vehicles Symposium, 2003. Proceedings. IEEE*, 2003.

[4] "Vehicle Hardware-In-the-Loop," [Online]. Available: https://www.tassinternational.com/VeHIL. [Accessed 18 08 2016].

[5] N. Hiblot, D. Gruyer, J.-S. Barreiro and B. Monnier, "Pro-SiVIC and ROADS. A Software suite for sensors simulation and virtual prototyping of ADAS," in *Proceedings of DSC*, 2010.

[6] "Chassis Dynamometer VULCAN," [Online]. Available: http://www.horiba.com/automotive-test-systems/products/mechatronic-systems/engine-test-systems/details/vulcan-emscd48-626/. [Accessed 18 08 2016].

[7] J. Petit, B. Stottelaar, M. Feiri and F. Kargl, "Remote Attacks on Automated Vehicles Sensors: Experiments on Camera and LiDAR," in *Black Hat Europe*, 2015.

[8] B. Steux, P. Coulombeau and C. Laurgeau, "RTmaps: a framework for prototyping automotive multi-sensor applications," in *Proc. Intelligent Vehicles Symposium*, 2000.

[9] I. Gat, M. Benady and A. Shashua, "A monocular vision advance warning system for the automotive aftermarket," in *SAE World Congress & Exhibition*, 2005.

[10] "MobilEye 5-series product page," [Online]. Available: https://www.mobileye.com/products/mobileye-5-series/. [Accessed 18 08 2016].

Index

About the Editors

Guillermo Payá Vayá obtained his Ing. degree from the School of Telecommunications Engineering, Universidad Politécnica de Valencia, Spain, in 2001. During 2001–2004, he was a member of the research group of Digital System Design, Universidad Politécnica de Valencia, where he worked on VLSI dedicated architecture design of signal and image processing algorithms using pipelining, retiming, and parallel processing techniques. In 2004, he joined the Department of Architectures and Systems at the Institute of Microelectronic Systems, Leibniz Universität Hannover, Germany, and received a Ph.D. degree in 2011. He is currently Junior Professor at the same Institute. His research interests include embedded computer architecture design for signal and image processing systems.

Holger Blume received his diploma in electrical engineering in 1992 at the University of Dortmund, Germany. In 1997 he achieved his Ph.D. with distinction from the University of Dortmund, Germany. Until 2008 he worked as a senior engineer and as an academic senior councilor at the Chair of Electrical Engineering and Computer Systems (EECS) of the RWTH Aachen University. In 2008 he got his postdoctoral lecture qualification. Holger has been Professor for "Architectures und Systems" at the Leibniz Universität Hannover, Germany, since July 2008 and runs the Institute of Microelectronic Systems. His present research includes algorithms and heterogeneous architectures for digital signal processing, design space exploration for such architectures as well as research on the corresponding modeling techniques.

About the Editors

CPSIA information can be obtained
at www.ICGtesting.com
Printed in the USA
BVOW07*1110080517
482744BV00001B/1/P

9 788793 519145